BuzzFeed

BRING ME!

THE TRAVEL-LOVER'S GUIDE TO THE WORLD'S MOST UNLIKELY DESTINATIONS, REMARKABLE EXPERIENCES, AND SPECTACULAR SIGHTS

LOUISE KHONG AND AYLA SMITH

RUNNING PRESS
PHILADELPHIA

Running Press
Hachette Book Group
1290 Avenue of the Americas, New York, NY 10104
www.runningpress.com
@Running_Press

Printed in China

First Edition: July 2022

Published by Running Press, an imprint of Perseus Books, LLC, a subsidiary of Hachette Book Group, Inc.
The Running Press name and logo is a trademark of the Hachette Book Group.

The Hachette Speakers Bureau provides a wide range of authors for speaking events.
To find out more, go to www.hachettespeakersbureau.com or call (866) 376-6591.

The publisher is not responsible for websites (or their content) that are not owned by the publisher.

Print book cover and interior design by Jay Fleckenstein & Joshua McDonnell
Interior and cover illustrations by Jay Fleckenstein, Michael Kilian, & Ivy Tai

Library of Congress Control Number: 2021951436

ISBNs: 978-0-7624-7494-3 (hardcover), 978-0-7624-7495-0 (ebook)

RRD-S

10 9 8 7 6 5 4 3 2 1

PHOTO CREDITS: Copyright Getty Images: p xxiii: RobHowarth, p xxviii R.M. Nunes, p 4: MARTIN BERNETTI, p 22: maylat, p 23: AnetteAndersen, p 32: TonyFeder, p 34: annaswe, p 40: Arnaud_Martinez, p 43: Bombaert, p 44: patrickds, p 48: pawopa3336, p 51: mikdam, p 57: Dendron, p 60: Tristan Barrington Photography, p 65: Anita Sagastegui, p 66: shalamov, p 67: ferrantraite, p 71: EunikaSopotnicka, p 72: TOSHIHARU ARAKAWA, p 73: cinoby, p 78: blueye, p 80: Simon Dannhauer, p 81: Wysiati, p 84: Thinkstock Images, p 87: tjalex, p 88: Elizabeth M. Ruggiero, p 90: AWelshLad, p 92: EGE APAYDIN, p 94: Olga Kaya, pp 100–101: ultramarinfoto, p 105: estivillml, p 106: CAHKT, p 107: wilpunt, p 111: flocu, p 113: dane-mo, p188: INTI OCON / Stringer, p 123: Jupiterimages, p 130: leungchopan, p 137: elkaphotos, p 139: lechatnoir, p 142: Anadolu Agency, p 144: Lisovskaya, p 149: Nicolas McComber, p 151: StockByM, p 157: Zhang Peng, © Shutterstock: P iv: Skreidzeleu, p xiv: Susan Montgomery, p xviii: Uwe Aranas, p 3: Peter de Kievith, p 8: Mrugas PHOTOgraphy, p 10: Elaine.Adoptante, p 12: Evgeniy Gorbunov, p 15: Mukkesh Photography, p 17: Guillermo Ossa, p 21: Steve Barze, p 23: Anton_Ivanov, p 28: Applepy, p 55: zcesty, p 69: elena_prosvirova, p 109: Anton Petrus, p 117: wayfarerlife, p 120: S-F, p 128: Juan Camilo Jaramillo, p 129: Julia700702, p 134: tipwam, p 147: LyndonOK, p 150: PixHound, p 155: Konmac, pp 168–169: Mel Gonzalez, p 170: Ludovic Farine, p 173: Pedro Quartin Graca,

CONTENTS

INTRODUCTION

Hello there, fellow traveler!

If the year 2020 taught us one thing (other than how to bake a good sourdough bread), it's to never take travel for granted. Suddenly, after the coronavirus pandemic struck, travel as we knew it was canceled, and we found ourselves pining for the weirdest things. Oh, what we'd have given to wait in a long security line, buy overpriced airport sandwiches, and get stuck in the middle seat between two strangers on a long-haul flight, if only it meant we could visit a destination that was anywhere but home.

Like many other grounded travelers, we started exploring our own backyards and discovered new corners of the world vicariously, spending hours scrolling Instagram as a way to temporarily keep wanderlust at bay. As a result, our never-ending bucket list grew longer and longer while our passports sat around collecting dust.

Now that travel is possible again, we're faced with a new problem: How can we possibly decide where to go and what to do? We've been reminded what a massive privilege it is to have the time, money, health, and freedom to visit a new city, state, country, or continent; to meet new people from all parts of the world; to learn about different cultures firsthand; to feel the rush of new experiences; to taste new flavors that you didn't know were possible. Knowing just how precious travel is, we feel even more pressure to make sure every trip is the Best Trip Ever and to make every second—and every dollar—count.

BuzzFeed launched Bring Me! in 2017 (a.k.a. the Before Times) as a way for people to discover unique things to do, places to eat, and sights to see around the world. And despite all the unexpected changes in the past couple of years, this mission still rings true. This book isn't meant to be a comprehensive destination guide that covers all the biggest tourist attractions in the world—you don't need us to tell you about the Eiffel Tower. Instead, this book is a collection of experiences that are a little more unique, unusual, or on the quirky side, experiences that are memorable, shareable, and worthy of a precious slot on your bucket list. Whether you're planning an upcoming trip or exploring the world from the comfort of home in your pj's, we hope you'll find something in this book to awe and inspire you.

These are the places around the world that will make you want to say, "Bring me!"

A NOTE FROM THE AUTHORS

The research for this book was finished in early 2021, when the pandemic was still far from over. We've tried our best to make sure the information in this book is up-to-date and accurate; however, it's possible some changes and closures may have occurred since this book's publication date. Please check local listings for updates before you visit the places in this book. Either way, we hope you'll find the information in these pages entertaining, interesting, or inspiring for your next trip.

HOW TO USE THIS BOOK

We're firm believers that the experiences you have are more important than the places you go, so that's why we've laid out this travel book a little differently. Instead of browsing by destination like a traditional destination guide, you'll be able to explore four categories of experiences: culture, nature, thrills, and taste. You can either flip through this book from front to back or jump straight to the section that speaks to your travel personality. If you're feeling lost, take the "What Type of Traveler Are You?" quiz on page xxvi, and we'll be glad to point you in the right direction.

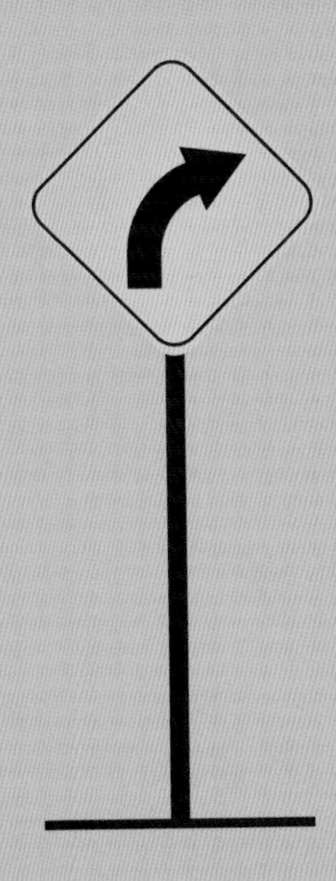

THE BRING ME! TRAVEL COMMANDMENTS

Just as there's no one right way to plan a trip, there's also no one right way to travel. But whether we're staying local or flying to the other side of the globe, these are the ten principles we always try to follow when we travel. You can follow them, too, or you can choose to ignore them, but either way, we hope they'll inspire you to think of your own personal travel commandments.

1. BE KIND TO THE ENVIRONMENT

It might sound obvious, but travel can have a devastating impact on the planet. So the more you can do to minimize your environmental impact while traveling, the better. Whether it's following the principles of Leave No Trace (page 106) when you're visiting the great outdoors, bringing your own water bottle to refill on your travels, or not leaving the lights running in your hotel room when you're not there (just 'cause you're not paying the electricity bill doesn't mean it "doesn't count"), there are countless ways to make small, conscious choices to decrease your footprint while you're exploring this beautiful Earth we all call home.

2. BE CURIOUS

You can spend months (years!) reading, watching, and soaking up information about a destination, but nothing compares to what you'll learn once you're actually there. Strike up a conversation with your taxi driver, speak to locals about their favorite spots, and ask your fellow travelers for their tips. Follow your curiosity—you never know where it might lead you.

3. DON'T BE A HATER

We get it: travel can be stressful, and it doesn't always bring out the best in us. But even in the most unfortunate situations, focusing on the good instead of the bad can go a long way. Negative energy radiates and won't just affect your trip—it'll also drag down the experience of people around you. It's totally fine to be annoyed every now and again, but have a quick vent, take a few deep breaths, and let it go. No one wants to travel with a hater.

4. SUPPORT LOCAL BUSINESSES

One of the best things you can do to support the economy and people of a place you visit is to shop locally. So take that extra step to research and see if the store, restaurant, or other place of business you're supporting is actually owned by locals. Even better, try to support small businesses that are owned by traditionally marginalized groups in the community.

5. GET COMFORTABLE WITH BEING UNCOMFORTABLE

And we don't just mean on a long-haul flight. Visiting new places, meeting new people, and trying new things can feel strange and scary at times, but stepping out of your comfort bubble and exposing yourself to different experiences is how you grow. Being able to travel and learn about other cultures through firsthand experiences is a huge privilege, so be thankful and make the most of it. If you wanted to be comfortable, you'd have stayed on the couch.

6. BE EMPATHETIC

One of the best parts of traveling is the people you meet. Wherever you go in the world, you're probably going to interact with people from all different cultures, whose life experiences vary drastically from yours. Leave your judgment at home with that fifth pair of shoes you *definitely* won't need and embrace the opportunity to better understand the experience of others.

7. EMBRACE THE UNEXPECTED

You can plan your entire trip to a tee, but some of the best moments you have while traveling are going to be completely unplanned and spontaneous. So don't be afraid to diverge from your itinerary every once in a while and say *yes* to something different if the opportunity presents itself, because these are usually the experiences that end up being the most memorable.

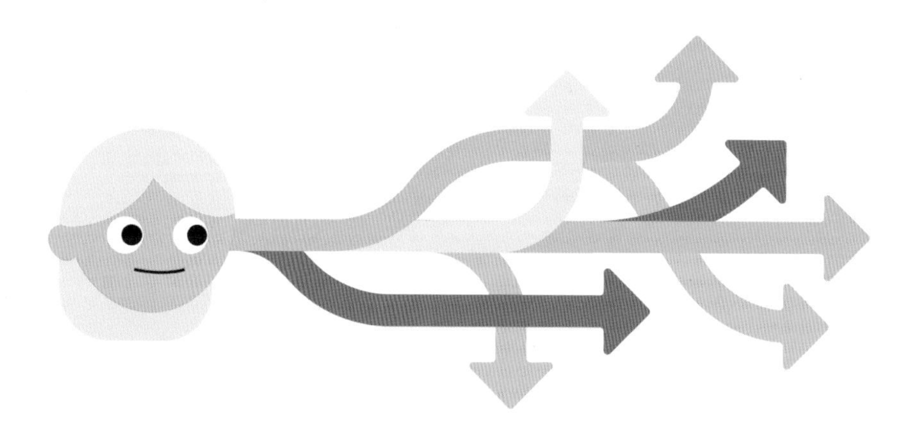

8. HEALTH ABOVE ALL ELSE

If there's one thing you *should* absolutely plan ahead for, it's your health. Make sure you're prepared for any personal or public health emergency. Travel insurance with health benefits is usually a good idea, as is researching the potential health hazards and medical care options at any place you're visiting. Learn how to ask for help in the local language and take note of the emergency phone numbers in each country you visit; remember, only a fraction of the world uses 911.

9. REMEMBER, IT'S NOT ALL ABOUT YOU

Sometimes it's easy to forget that nearly every place you visit—no matter how much it feels like it exists to cater to tourism—is also somebody else's home. When traveling, remember that not everyone is here to make sure you have an enjoyable trip and not everyone speaks the same language you do. Being respectful of other people's spaces and communities will ultimately lead to a better travel experience.

10. DO IT FOR THE MEMORIES

Don't get us wrong, we're all for snapping pics and documenting travels, but it can be easy to get so caught up trying to take the perfect shot that you forget to really take in the experience. You didn't travel all that way to look at the world through *another* screen, so remember to put your camera down every now and then and enjoy the moment.

PREPARING FOR YOUR TRIP

WHAT TO PACK

Are you an overpacker or an underpacker? Team Suitcase or Team Backpack? A roller or folder? Or maybe you're someone who just throws it all in and hopes for the best? Love it or hate it, packing is an inevitable part of traveling, and while it helps to be prepared, thankfully these days you'll be able to pick up most essentials in any place you visit (so don't panic if you forget your toothbrush). Clothes will probably make up the majority of your luggage space, so the biggest factor to consider is the weather: check the monthly average temperature and rainfall at your destination(s) so you can plan your wardrobe accordingly.

TIPS

- Suitcases can be great for rolling around big cities or settling into a resort, but the wheels don't always fare well on cobblestone or dirt roads. If you're planning to move around a lot, a backpack might be more convenient—just as long as you're able to carry it!
- Opt for light layers rather than big, bulky items of clothing, especially if you're packing for multiple seasons or climates. Throw in a travel-size laundry detergent, too.
- Since you'll probably be eating on the go more than you usually would, pack a reusable metal straw and a spork so you don't have to eat with (and throw away) single-use plastic cutlery all the time.
- Invest in some reusable leak-proof travel-size bottles and fill them up with your favorite bath products. It's cheaper in the long run *and* better for the environment. Also, as cute as the mini bottles of hotel toiletries can be, try to avoid using them if you can: all that teeny-tiny packaging creates a lot of waste.
- Make a mini first-aid kit with some basics like bandages, pain killers, antiseptic cream, and any medication you'll need while you're away.
- Airline rules for carry-on bag sizes and liquid restrictions can vary from place to place, so be sure to check and adhere to local regulations if you're flying internationally.

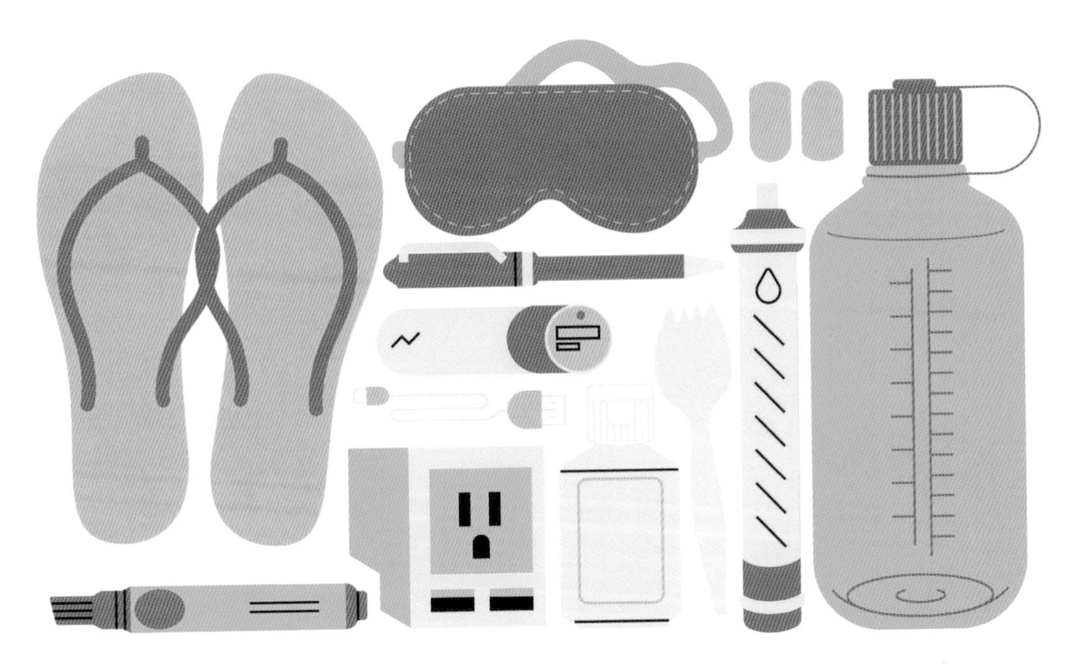

Along with the obvious things, like your passport and toothbrush, here are ten essential items that'll more than earn their spot in your bag:

1. a spork, for impromptu street snacking
2. hand sanitizer, obviously
3. a laundry pen, for any spills
4. flip-flops, in case you need to use a public shower
5. an eye mask, for optimal napping conditions no matter the time or place
6. a set of earplugs; see above
7. a portable phone charger, so you're never out of juice when you encounter that perfect photo
8. a pen, because there's usually some form or another to fill out
9. a universal adapter, so you don't need to buy a different one every time you visit a new country
10. a water-filter straw and refillable bottle, so you're never without fresh H_2O

WHERE TO STAY

Accommodation is so much more than just a place to rest your head at night: it can take your trip from good to great. And "great" doesn't have to mean a swanky five-star resort. While there's nothing wrong with a little luxury every now and again, it really all depends on the type of trip you're after. Fancy hotels might be the way to go if you're planning a romantic vacation with your significant other, but if you're headed on a solo adventure, you might have more fun staying at a budget hostel where you can meet other travelers. Or if you're cruising with the whole crew, you might want to opt for a rental with more space so you can all bunk together.

TIPS

- Go beyond the city center. It's easy to default to a hotel smack-bang in the middle of downtown—and sometimes that's perfect! But it's not your only option. Staying in fringe neighborhoods (where locals actually live) can help give you a more authentic feel of the city.
- Repeat after us: Read the reviews! Learning about the experience of past guests is the best way to get an idea of what the property is really like. Of course, you can take them with a grain of salt—just use your judgment.
- Nightly rates change often, so always compare prices on different booking websites before locking in your stay. Booking directly with the property can sometimes be cheaper than going through a third party.
- If you're staying at a hotel, ask if you can opt out of daily linen and towel services. Using fresh sheets and towels every day comes at a huge environmental cost, and saying no to these services is a simple way to reduce your impact.

If you're looking for a different kind of lodging experience (read: a really cool one), we've rounded up some of the more interesting and fantastical places around the world to rest your head when you're traveling.

CLIMB UP TO A TWELVE-HUNDRED-FOOT-HIGH SKYLODGE

Skylodge Adventure Suites—Cusco, Peru

This suspended lodge gives a whole new meaning to the phrase "climb into bed"—you literally have to scale twelve hundred feet (366 meters) up the side of a mountain before you can hit the sack. The Via Ferrata route is made for beginners, with secured cables and ladders, and the transparent capsules offer panoramic breathtaking views of the Sacred Valley below. You can even enjoy a Peruvian feast atop your capsule and, when it's time to leave, zip-line back to solid ground.

WAKE UP IN THE FUTURE AT A JAPANESE CAPSULE HOTEL

Nine Hours—Various Locations, Japan

Capsule hotels aren't a new thing—they were first invented in the '70s as a place for Japanese salarymen to rest for the night after a long day in the office (and an evening at *izakayas*)—but in recent years they've grown in popularity, both in Japan and abroad. It's kinda like bunking in a hostel dorm, except with more privacy since you have your own private sleep pod. Nine Hours puts a futuristic spin on the concept, with slick modern designs that'll make you feel like you're staying in a spaceship.

DREAM AWAY IN A SPHERICAL TREE HOUSE

Free Spirit Spheres—Vancouver Island, Canada

Spending the night hanging in a spherical tree house surrounded by forest sounds like the most magical thing ever—and on Vancouver Island, you can do just that. There are only three spheres on this secluded property, each equipped with big, round windows (for optimal nature viewing) and accessible via a spiral staircase.

PLAY ALL NIGHT AT A BOARD-GAME HOUSE

Great Escape Lakeside—Groveland, Florida, United States

You probably won't get much sleeping done at this house in Florida—it's basically a massive board game. There are thirteen bedrooms, each with a different game theme, from Clue, to Operation, to Scrabble. If that's not enough, there's also a *Jumanji*-themed movie theater, a studio where you can play TV game shows like *Family Feud*, a laser maze, a pool, and So. Much. More.

GET SOME FIRST-CLASS SHUT-EYE IN A CONVERTED PLANE SUITE

Costa Verde 727 Fuselage Home—Quepos, Costa Rica

Sleeping on an old airplane might not sound like the most comfortable thing, but this salvaged 1965 Boeing 727 airframe has been refurbished into a plush two-bedroom suite, complete with private bathrooms, air-conditioning, a kitchenette, and an ocean-view terrace. Set in the middle of the Costa Rican jungle and hugged by plants, a night here is truly unforgettable.

STAY THE NIGHT IN AN UNDERWATER HOTEL ROOM

Manta Resort—Pemba Island, Tanzania

Live out your *Little Mermaid* fantasies at this hotel room under the sea. Anchored to the ocean floor, the floating suite has three levels: the underwater bedroom with big perfect-for-fish-spotting windows, a lounge and bathroom, and a rooftop sunbathing/stargazing bed. You'll find it on Zanzibar's remote Pemba Island.

READ YOURSELF TO SLEEP AT A LIBRARY B&B

Gladstone's Library—Flintshire, Wales

Contain your excitement, bibliophiles: you can sleep overnight at this library in the United Kingdom. With twenty-three boutique rooms, it's the perfect spot to think, write, or read; the library collection contains more than 150,000 items, and guests have access to the reading rooms until 10:00 p.m.

WARM UP IN BED AT A SNOW HOTEL

SnowHotel—Kemi, Finland

Each year in Lapland, Finland, this impressive hotel is carved out of snow and ice. Every room is given a unique design and decorated with ice art. As for the bed, that's see-through and frozen, too, and the room temperature gets down to 32 degrees Fahrenheit (0 degrees Celsius), but don't worry: the plush sleeping bags will keep you nice and toasty. Find a cooler place to sleep—we dare you.

SPEND A NIGHT IN A SILO

SiloStay—Little River, New Zealand

A little under an hour from Christchurch in the small town of Little River, this eco-conscious motel features nine silos that have been converted into multilevel apartments with kitchenettes, private bathrooms, bedrooms, and even balconies. It doesn't get much quirkier.

SLEEP IN A HOTEL MADE OUT OF SALT

Palacio de Sal—Colchani, Bolivia

If a visit to the world's largest salt flat, Salar de Uyuni, in Bolivia, wasn't already on your agenda, this luxe salt hotel might just zoom it to the top of your bucket list. More than one million salt bricks were used to construct the building, which sits at an altitude of 12,140 feet (3,700 meters), and some of the furniture is made from salt, too.

GETTING AROUND

You might think of transport as a means to an end when you're traveling, but it can be just as memorable of an experience as anything else on your trip. If you plan it right, this is when you get to quietly pause, sit back, and reflect on your travels (not to mention enjoy the views outside). Sure, navigating transport can often be the most stressful part of any trip—but the more time and effort you put in researching this part in advance, the less hassle you'll have to deal with when it comes to getting around safely and smoothly.

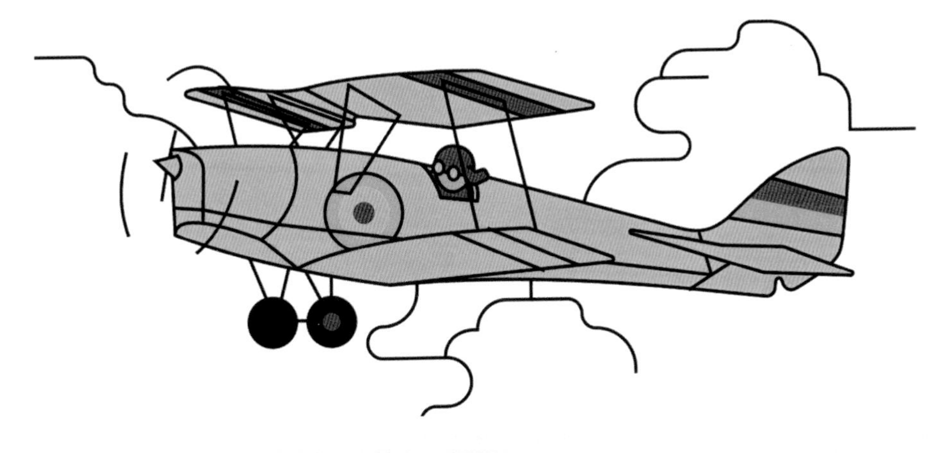

TIPS

- If you're taking a taxi without a working meter, negotiate the price with the driver up front. You can even write it down on a piece of paper that you both acknowledge.

- Always travel with licensed taxis or transport services. Watch out for unmarked cars with no meters or visible driver's license inside.

- Use a taxi app or rideshare service like Uber for additional peace of mind. You can follow along on a map to make sure the driver is taking the best route.

- If you plan to get around using public transportation, research the public transport system before you travel. Sometimes you can save money by buying a day-tripper pass online in advance, and many cities have handy apps to help you navigate their metro systems.

- Download maps to your phone. If you're a paper-map or guidebook kind of traveler, consider hiding your map inside another book or finding a cover for your guidebook so you look less conspicuously like a tourist.

Whether you're looking for a more interesting way to get from point A to B or a round-trip experience, here are some of the more unique transport options we've found.

TRAVEL THROUGH REDWOOD FORESTS ON THE SKUNK TRAIN

Skunk Train—Fort Bragg and Willits, California, United States

Sit back and enjoy the views on this historic train that winds through the redwood forests of Northern California. The California Western Railroad—commonly known as the Skunk—has a long history, but the views are as fresh and majestic as ever. You can also cycle along the Redwood Route on a railbike.

BIKE AND DRINK BEERS IN AMSTERDAM

Amsterdam Beer Bike—Amsterdam, Netherlands

Combine exercise, sightseeing, and beer (or prosecco) with the Amsterdam Beer Bike. It's a multiperson bike that's also a bar, with a roof, a sound system, and a driver to steer you in the right direction. Pedal away as you all make your way through the streets of Amsterdam and through the five gallons (twenty liters) of beer available per bike.

TAKE A TRAIN THAT TRAVELS OVER WATER

Rameswaram Express—Mandapam and Pamban Island, India

The Pamban Bridge is a railway bridge that connects the town of Mandapam in mainland India to Pamban Island (also known as Rameswaram Island). The bridge runs incredibly low and close to the water, so let's just say you wouldn't want to cross it in a storm. Hop aboard the Rameswaram Express train to cross the bridge to visit Rameswaram, which is considered one of the most sacred places in India.

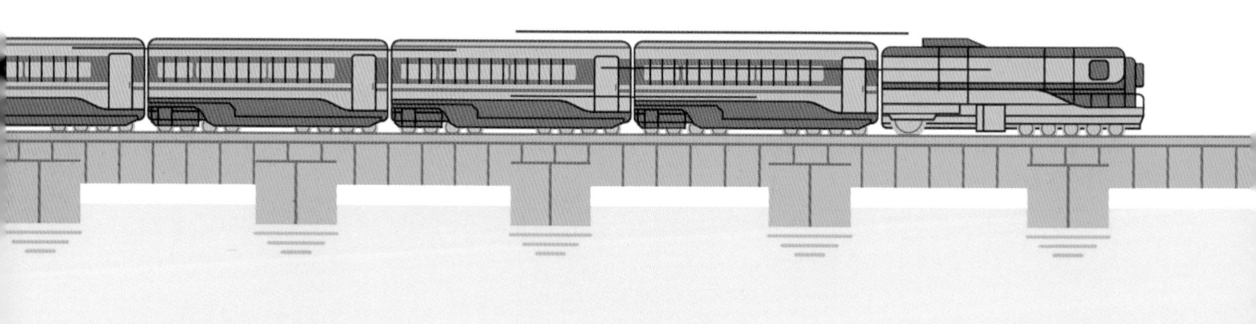

EXPERIENCE THE WORLD'S FIRST "FLOATING TRAM"

Schwebebahn—Wuppertal, Germany

Wuppertal is one of Germany's greenest cities, known for its parks, woods, and gardens. But it's most famous for being home to the world's first suspension railway, the Wuppertaler Schwebebahn. Rather than run along the top of the monorail, the train hangs below it, where it floats through the city and over the river, carrying visitors and locals alike.

FLOAT AROUND COPENHAGEN ON A PICNIC BOAT

GoBoat—Copenhagen, Denmark

For a unique experience in Copenhagen, rent a picnic boat with GoBoat and go for a peaceful cruise around the harbor. These sun-powered electric boats have wooden picnic tables in the center and can fit up to eight people, so they're perfect for a quick, romantic couples cruise, a family outing, or a six-hour boozy (but not *too* boozy) picnic with your crew. You can pack a picnic or buy organic snacks and drinks before you set out.

RIDE ABOARD THE CAMBODIAN BAMBOO TRAIN

Bamboo Train—Battambang, Cambodia

The Bamboo Train is a unique transport experience in the village of O Dambong II, on the outskirts of Battambang, Cambodia. Known by locals as the "norry," it's a small bamboo platform with a motor attached that runs across the old railway tracks. If you ride just before sundown, you might be lucky enough to catch a beautiful sunset over rice paddies. Don't forget the beers!

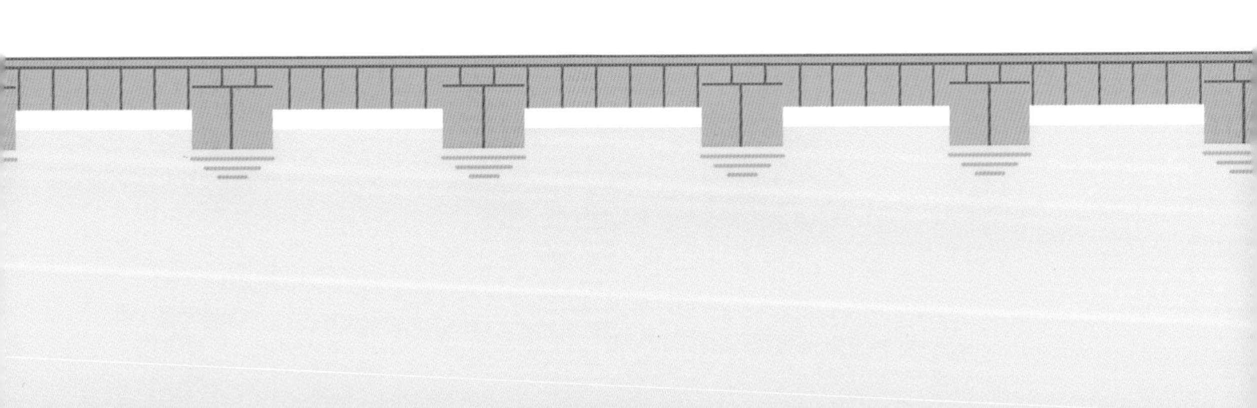

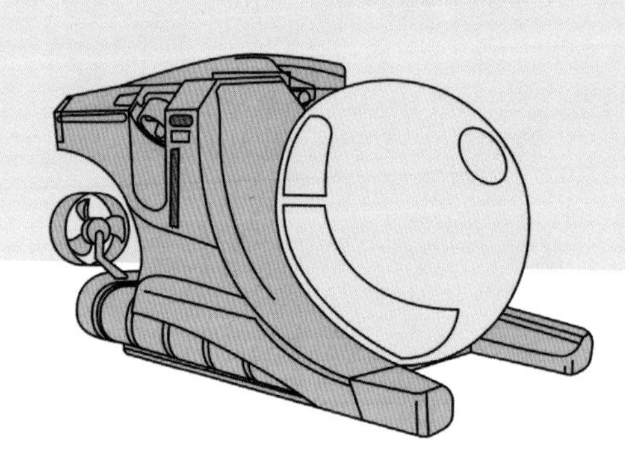

CHARTER YOUR OWN SUBMARINE

Charter a Sub—Various Locations

Luxury travelers, did you know you can charter a submarine that will take you deeper than is possible scuba diving? Whether it's exploring wreckages in Malta and reefs in Italy, or getting up close to marine life in Antarctica, this is an otherworldly experience that most of us can only dream of. You just need to BYO superyacht.

SLEEP OVERNIGHT IN A LUXURY BUS

Cabin—Los Angeles and San Francisco, California, United States

For a luxury (but more affordable than first-class) alternative to flying between LA and Francisco, book a trip with Cabin. This overnight bus lets you skip the airport and spend your trip sleeping in a cozy bed in a private cabin. You can stash your carry-on in a cubby, charge your phone, and get a good night's rest before waking up at 7:00 a.m. in your destination. Best of all, their Cabin Cloud bump-canceling technology means a smooth ride with no road turbulence.

FLY IN A VINTAGE AIRPLANE

Classic Wings—Duxford, England

If you want to relive the golden age of flying in a vintage aircraft, check out Classic Wings at the Imperial War Museum in Duxford. Don your helmet and goggles and take a flight in a Tiger Moth, or go for a longer scenic ride over Cambridge or London in a Dragon Rapide. You can even take a victory roll in a Spitfire, the iconic fighter aircraft that took its maiden flight in the late 1930s at this same airbase.

WIND THROUGH THE CANADIAN ROCKIES IN A GLASS-ROOFED TRAIN

Rocky Mountaineer—Vancouver, Canada

The Rocky Mountaineer is a luxury train experience that lets you travel in a glass-domed coach, giving you panoramic views of the most amazing scenery of western Canada and the American Southwest while you enjoy West Coast–inspired cuisine. The trains run in the daytime (because they're all about the views), but packages involve overnight stays in a hotel at each of your stops. Most trips start in Vancouver, Denver, or Las Vegas, and packages run from one to twelve nights.

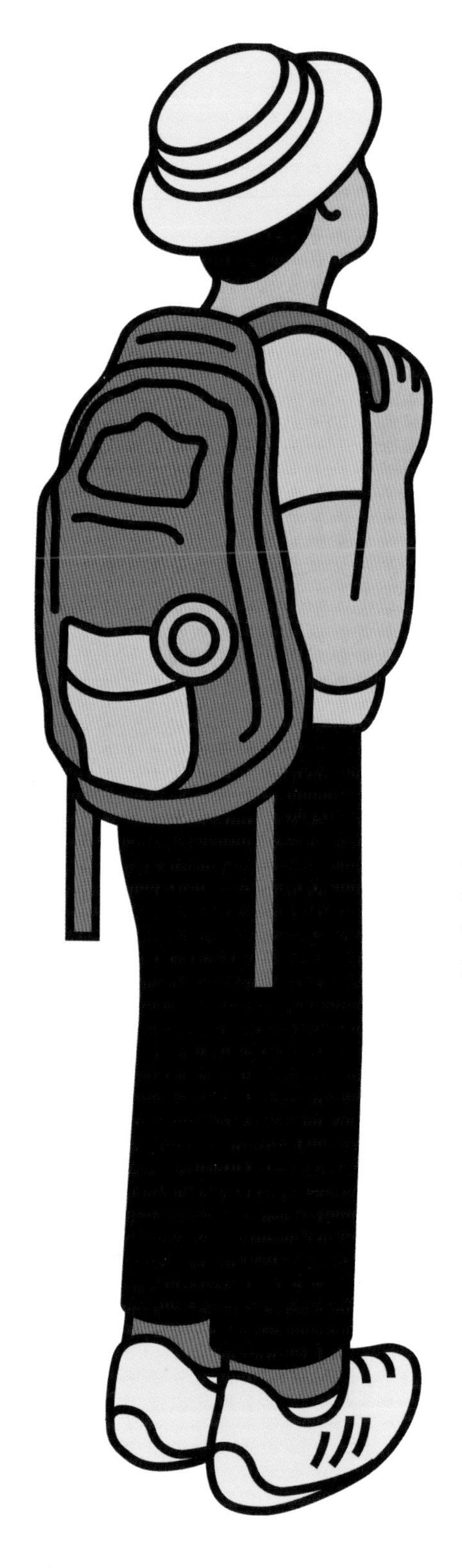

STAYING SAFE

As exciting as travel can be, it can also be daunting, and sometimes even dangerous. The world is full of unknowns, and no activity is 100 percent safe—but that shouldn't stop you from getting out there and exploring. We're all for stepping out of our comfort zones when we travel and we hope you are, too, but here are nine practical tips to help you do so safely.

TIPS

- **Research well and research often.** It's important to do your own in-depth research on any destination or experience on your itinerary. Consult the US State Department website for travel advisories and read reviews on trusted booking sites and travel forums before you make plans. Then continue checking local tourism or government websites for any warnings or updates.

- **Make a copy of your travel documents.** Create a digital copy of your passport, driver's license, any visas, flight details, and so forth, to keep in your email (or in the cloud or on a USB), and print out a copy that you can carry on you, in case anything happens to the originals.

- **Share your travel plans with a friend or family member back home.** It's helpful for someone to know where you are or where you're meant to be, especially in an emergency. Plus, they can follow along and live vicariously through you!

- **Carry some cold hard cash with you.** Electronic payments are so prevalent in the United States that you couldn't be blamed for forgetting cash is even a thing. But that's not the case everywhere. Keeping a bit of cash on you can get you out of a sticky situation, like if the only ATM at the airport doesn't accept foreign bank cards or if you're *really* hungry and need a snack from a street-food vendor pronto.

- **Take health precautions to protect yourself (and those around you).** If it's been awhile since your last checkup, it could be a good idea to visit a doctor before you travel. Research potential health risks at your destinations and prepare by getting any recommended vaccinations. And don't forget to take your new best friend—hand sanitizer.

- **Bring a backup credit or debit card.** Store it in a separate bag or pocket from the rest of your money.

- **Figure out what level of risk you're comfortable with.** We take risks every single day, and the same goes when we travel. But when you're considering things like hiking the rim of an active volcano or scuba diving in uncharted waters, the potential dangers are right up in your face. Just because other people might be willing to take on certain risks, that doesn't mean you have to. Don't be pressured into doing anything you don't want to do and stay true to what feels right for you.

- **Consider travel insurance.** It's not for everyone, but if you think you'd benefit from a little peace of mind, it might be worth it. Before you take out any policy, be sure to read the fine print carefully (yes, all of it) so you know what is and isn't covered.

- **Trust your gut.** You're more instinctive than you may realize. Listen to that old intuition of yours. If a situation doesn't feel right, get out of there.

QUIZ: WHAT TYPE OF TRAVELER ARE YOU?

Are you thinking about planning your next trip, but you're not sure where to start? Answer a few brief questions, and we'll reveal what type of traveler you are and where you should start in this book. It's science!

How would your travel companion describe you?

A. Curious
B. Outdoorsy
C. Daring
D. Hungry

What's the first thing you pack in your suitcase?

A. A good book
B. A swimsuit
C. Hiking boots
D. A spork

Where are you sleeping?

A. A boutique hotel
B. A tree house
C. A tent
D. A local rental

Describe your dream trip in one word:

A. Fascinating
B. Relaxing
C. Exciting
D. Delicious

What's your dream destination?

A. Italy
B. Costa Rica
C. South Africa
D. Thailand

Which activity is at the top of your list?

A. People-watching
B. Stargazing
C. Zip-lining
D. Street-food tour

Pick a souvenir to bring home with you:

A. Local artwork
B. A hat
C. I'm not coming home
D. Snacks

IF YOU GOT MOSTLY A'S:

You're a Culture Lover!

You're a cultural sponge who's ready to soak up everything a destination has to offer. You're truly fascinated by the world, and for you travel is a way to learn about the people and places that make it special. You can spend hours marveling at museums, exploring quirky towns, or simply people-watching on a busy street corner. You're a curious soul, and you won't stop until you've seen it all.

Where to start: Flip to page 1 for out-of-the-ordinary travel experiences that'll have you packing your suitcase ASAP.

IF YOU GOT MOSTLY B'S:

You're a Nature Wanderer!

You prefer to spend your trip outside, and you're happiest when you're surrounded by nature. Exploring the wilderness, spending time with the local wildlife, and relaxing on as many beaches as possible—this is what travel is all about. You love wandering through lush green forests or gazing up at wide-open skies, and even if you're traveling in the city, you still seek out the hidden gardens and pockets of greenery.

Where to start: Flip to page 42 to find the best places and experiences for nature lovers.

IF YOU GOT MOSTLY C'S:

You're a Thrill Seeker!

You're a daredevil who's always on the lookout for your next adventure. You're confident and you believe some of the best experiences come from taking risks, but you're not out to prove anything to anyone but yourself. You're naturally drawn to activities that get your blood pumping, from touring haunted cities to cage diving with crocodiles. You're up for anything that makes you feel more alive.

Where to start: Flip to page 91 for experiences guaranteed to give you an adrenaline rush.

IF YOU GOT MOSTLY D'S:

You're a Taste Tripper!

You're in it for the food. You're quite happy planning your entire trip around where you'll be eating, and you want to sample absolutely everything a travel destination has to offer. You like seeking out the most unique places to eat, and you're an adventurous eater, so you'll order one of everything on the menu. Your ideal travel partner is someone who'll try everything with you—as long as they'll let you have some of what's on their plate, too.

Where to start: Flip to page 138 to find unique foods and eating experiences to inspire your next trip.

PART ONE

CULTURE LOVER

The ability to experience another culture firsthand is one of the greatest privileges travel affords. Not only is learning about a place and its people through buildings, galleries, museums, and festivals incredibly fun, but it also makes your trip—nay, your life—far richer. The following ideas will take you off the well-trodden path and down a road dotted with unusual sites and activities you've never heard of.

Be inspired by some of the world's most unique architecture and take in contemporary artworks set among breathtaking natural landscapes. Illuminate your travels—and your camera roll—as you wander the world's most colorful places and leave the major cities to get to know the locals in smaller, quirkier towns. Whether you're visiting a place for the first time or the fifteenth, there are always new layers to peel back, new stories to uncover, new people to meet, and new perspectives to understand.

EXTRAORDINARY ART

There's plenty of art to see in the world's great galleries and museums, and we don't need to tell you where they are. But there's so much more to find. From contemporary galleries and art islands to outdoor installations worthy of your attention, we've rounded up just some of the extraordinary art waiting to be discovered around the world.

DAYDREAM BENEATH A BILLOWING SKY NET

Bending Arc—St. Petersburg, Florida, United States

Art you can enjoy lying down, and for free? Sign us up. This lightweight aerial net sculpture is constructed of more than a million knots and a whole lot of rope, and gently floats above Pier Park in St. Petersburg, Florida. Lounging on the grass and watching the installation billow and dance in the wind is truly a meditative experience. At night it is lit up in violet and magenta lights and seems to ripple like the aurora borealis across the sky. The park has significance in the Civil Rights Movement, which inspired artist Janet Echelman to name her work after this Martin Luther King Jr. quote: "The arc of the moral universe is long, but it bends toward justice."

WANDER THE "ISLAND OF DREAMS"

Djerbahood—Djerba, Tunisia

Djerba, Tunisia's "island of dreams," is known for El Ghriba, Africa's oldest synagogue. But lesser known is that in 2014, more than a hundred street artists from around the world turned one of the island's villages, Erriadh, into an open-air museum. Visitors can wander through quiet alleyways and admire the murals and pops of color that decorate the buildings, along with colorful flowers and palms. The art extends out beyond the village, too, adorning an abandoned prison and an eighteenth-century palace. It is a dreamy place.

EXPLORE EUROPE'S LARGEST ART PARK

Nikola-Lenivets—Kaluga Oblast, Russia

Nikola-Lenivets is an art park in the Kaluga region, southwest of Moscow. Situated in the green Ugra National Park, it's built around a village and features dozens of incredible landscape installations and art objects from Russian and international artists. Visit in July for the Archstoyanie—unofficially known as "Russia's Burning Man"—to admire the permanent land art and architecture, along with new installations made from natural materials that are all ceremoniously burned at the end of the festival.

GAZE UP AT ROTTERDAM'S "HORN OF PLENTY"

The Markthal—Rotterdam, Netherlands

Known as the "Sistine Chapel of Rotterdam," the Horn of Plenty is a massive mural artwork adorning the arched ceiling of the Markthal, the first covered market in the Netherlands. It's a vibrant depiction of abundance: fruits and flowers, plants and vegetables, bees and butterflies. The piece was created by Arno Coenen with the help of a team that included Pixar Animation Studios, which transformed the artwork into a file that could be printed onto four thousand individual panels. Some call it the largest artwork in the world, but that's up for debate. In any case, it's big, and it's a bright and delightful piece to take in while you browse the fresh food market.

FIVE PLACES TO SEE INCREDIBLE STREET ART AROUND THE WORLD

1. **IPOH, MALAYSIA**
   Wander the Old Town of Ipoh to find iconic murals by Lithuanian artist Ernest Zacharevic or the New Town's Mural Arts Lane to discover works by local artist Eric Lai.

2. **BRISTOL, ENGLAND**
   Bristol is world famous for its street art, and Nelson Street is where you'll find some of the best. But there's outdoor art all over the city, including works by Banksy, who (supposedly) hails from there.

3. **BALTIMORE, MARYLAND, UNITED STATES**
   Station North Arts & Entertainment District features amazing art walks and murals from street artists around the world, including the Open Walls Baltimore project curated by the artist Gaia.

4. **JOHANNESBURG, SOUTH AFRICA**
   Joburg, like Cape Town, has a vibrant street-art culture. Start around Arts on Main in the Maboneng Precinct, a cultural hub with plenty of art to explore both indoors and outdoors.

5. **VALPARAÍSO, CHILE**
   This colorful port city is the best place in Chile to see street art, in part because the government actually encourages it. From Concepción Hill to the Pablo Neruda dedications at the Open Sky Museum of Bellavista Hill, there is an abundance of colorful murals, paintings, and mosaics to see.

TAKE A COASTAL CLIFF-TOP ART WALK

Sculpture by the Sea—Sydney, Australia

Australia's iconic beaches are beautiful year-round, but once a year a coastal walk starting at Sydney's Bondi Beach is transformed into a spectacular outdoor sculpture exhibition. More than a hundred mind-bending, unique, and sometimes humorous creations pop up along the sand, on the rocks, and on the cliff tops, and visitors can stroll for a mile taking in both art and incredible views at once. The event has been so popular there's now also a west coast edition at Cottesloe Beach in Perth (which has epic sunsets too, just FYI).

WAVE HELLO TO A GIANT HAND IN THE DESERT

Mano del Desierto—Antofagasta, Chile

Deep in the Atacama Desert in Chile, a giant sculpture of a human hand by artist Mario Irarrázabal springs up from the sand. You can find it about a forty-five-mile drive along the Pan-American Highway from the city of Antofagasta, a strange and beautiful sight in an otherwise barren landscape. Spend some time in its presence and reflect on your own tininess. Or, you know, just take a photo. Lefties, rejoice; it's a left hand, so you are represented here. But for everyone else, you can find Irarrázabal's right-hand counterpart emerging from a beach in Uruguay.

NORWEGIAN CAVIAR, BUT MAKE IT ART

KaviarFactory—Nordland, Norway

Far up in northern Norway on Henningsvær island, in the archipelago of Lofoten, is an off-the-grid gallery in a former caviar factory. You might not find fancy fish roe in this remote seaside building anymore, but you *will* find an impressive collection of Norwegian and international contemporary art. It's also conveniently situated close to *Skulpturlandskap*, a sculpture trail and Artscape Nordland project that spans thirty-four municipalities in Nordland County.

GO ART ISLAND HOPPING IN JAPAN

Benesse Art Site, Naoshima—Kagawa, Japan

The Benesse Art Site is a collective of art museums and other installations across three islands in Kagawa and Okayama Prefectures in Japan. Naoshima is the most popular of the islands, and it's here you'll find the famous seaside yellow-spotted pumpkin by Yayoi Kusama. Elsewhere on Naoshima you can enjoy Claude Monet and James Turrell on permanent display at the underground Chichu Art Museum.

Travel Tip: Stay an extra day or two and explore the lesser-known galleries and art houses on nearby Teshima and Inujima islands.

DISCOVER CONTEMPORARY ART IN SHANGHAI

M50 Creative Park—Shanghai, China

M50 is a cluster of galleries, private art studios, cafés, and outdoor art set among restored textile factory buildings at 50 Moganshan Lu in Shanghai. Galleries and exhibitions pop up and then disappear, but there's always an abundance of cool things to see here if you know where to look. It pays to know in advance which places you want to visit, but if you're feeling spontaneous you can also just wander around to check out the street art and pop into random places you find along the way.

LISTEN TO RAIN ON A TIN ROOF

Cloud House—Springfield, Missouri, United States

What's more calming than listening to the patter of rain falling gently on a tin roof? Set in Farmer's Park in Springfield, Missouri, Cloud House is an installation by Matthew Mazzotta where you can do just that. It's a tiny wooden house with a marshmallow whip–white cloud sculpture set above it. The structure is designed to harvest rainwater, which the cloud automatically "rains" back down onto the roof when triggered by the rocking chairs that you can sit on inside the house. The raindrops roll back down to water edible plants in the windowsills, and the gentle cycle invites you to reflect on the natural systems that provide us with the food we eat. Plus, it's really relaxing.

UNUSUAL ARCHITECTURE

Contrary to popular belief, we'd argue it's not actually hip to be square. Forget boring old rectangular-shaped buildings and discover some of the world's most interesting structures, from trippy, warped shop fronts to shoe-shaped churches.

STROLL THROUGH A GIANT CELL

Atomium—Brussels, Belgium

If you find yourself in Belgium's capital, it's basically impossible to miss this giant chrome structure that towers 335 feet (102 meters). Created for the Brussels World's Fair back in 1958, the Atomium is shaped like an iron-unit cell (think of a cube made of rods and spheres resting on one corner) magnified a whopping 165 billion times. These days you can explore the spheres and stroll through the tunnels, but even if you don't make it inside, the view from the ground is arguably even more impressive. And while, yes, this site is touristy, it is definitely worth your time!

VISIT THE WORLD'S LARGEST PIECE OF POTTERY

Casa Terracota—Villa de Leyva, Colombia

Sure, handmade ceramic mugs and bowls are cool, but have you ever imagined a life-size house built entirely out of clay? Located to the northeast of Bogotá, that's essentially what this fairy tale–esque structure is: one massive piece of pottery to call home. It's the brainchild of architect Octavio Mendoza Morales, who sought to use the four elements—earth, fire, water, air—to create a habitable structure where the environment, art, and architecture combine.

MARVEL AT THE CROOKED BUILDING

Krzywy Domek—Sopot, Poland

This trippy building looks like it's been plucked from a scene in *Alice in Wonderland* and plonked down in the middle of an otherwise ordinary-looking street in the seaside town of Sopot, Poland. The warped walls, windows, and roofs were inspired by Polish illustrator Jan Szancer's sketches. It's part of the Rezydent shopping center, and inside you'll find an array of boutiques, food spots, offices, and other businesses. But here, it's more about what's on the outside that counts (for once).

ADMIRE A DRAGON-WRAPPED TEMPLE

Wat Samphran—Khlong Mai, Thailand

Thailand has thousands of stunning temples, but we're going out on a limb to say this one is the most unique. About twenty-five miles west of Bangkok, you'll find a pink tower reaching seventeen stories into the air with a mythical dragon coiled around it. Visitors have reported being able to climb through the belly of the dragon, with incredible views across Thailand's capital for those who make it all the way to the top.

Good to Know: While churches, temples, and mosques can be inspiring to see while traveling, they're places of worship, first and foremost. So remember to be respectful when visiting any religious or sacred site, read (and obey) the signs, and be mindful of local customs.

SEE THE WORLD'S BIGGEST MUD-BUILT STRUCTURE

The Great Mosque—Djenné, Mali

About 220 miles southwest of Timbuktu, perched on the banks of the Bani River, is Djenné, an ancient trading city and home of the largest structure built from mud in the world. Made mostly from sun-dried mud bricks, the Great Mosque is a respected example of Sahelian and Sudanese architecture that's believed to date all the way back to the thirteenth or fourteenth century. The impressive building you'll see today isn't the original—it was reconstructed in 1907, and, due to the area's harsh climate (think long hot spells followed by heavy rainfall, which isn't great for the mud bricks), the mosque has to be replastered with mud every year. This task has been turned into an annual festival called Crepissage de la Grand Mosquée, in which locals help complete the refurbishment.

SLEEP IN THE BELLY OF A GIANT SNAKE

Quetzalcoatl's Nest—Naucalpan, Mexico

About a forty-minute drive from Mexico City, in the forested town of Naucalpan, rests an enormous snake-shaped structure that's truly a sight to see. It was designed by architect Javier Senosiain, whose vision was to create a building that slotted into the landscape and promoted harmony between humans and nature. In true snake fashion, the building features many curvy lines and colorful patterns and is home to ten unique apartments—including one that's available for rent on Airbnb! It lives within Parque Quetzalcóatl, named after the Mesoamerican Feathered Serpent god, a huge property with plenty to explore, such as a botanical garden in a stained-glass snail shell, a grassy amphitheater, crystal-filled caves, interesting sculptures, and more. While you're in Naucalpan, be sure to also check out the surreal seashell house, a residential property Senosiain designed for a local family seeking a home that felt more in touch with nature.

HAVE A PICNIC NEAR A VERY BIG BASKET

Basket Building—Newark, Ohio, United States

Constructed in 1997, this unique building was once the head office of Longaberger, a company that made—you guessed it—woven baskets. It's seven stories tall, spans 180,000 square feet (16,723 square meters), and is thought to have cost around $30 million to build back in the day. Oh, and let's not forget the handles that weigh seventy-five tons! While there were plans to turn the building into a luxury hotel, it currently lies empty, so there's not a ton to do except marvel at the very big basket from the outside (and maybe have a picnic on the grass to honor what once was).

BE WOWED BY MUSICAL BUILDINGS

Piano House—Huainan, China

Music lovers, we've found your dream home! This impressive building in Huainan, in the north-central Anhui Province of China, is shaped like a huge black piano and features a striking transparent violin made of glass. The building itself supposedly functions as an urban planning center, but according to visitors it's often closed. For those who are just after a photo op (a.k.a. pretty much everyone), though, that doesn't really matter.

VIEW REYKJAVÍK FROM THE TOP OF AN ICONIC CHURCH

Hallgrímskirkja—Reykjavík, Iceland

Okay, so this church isn't exactly a hidden gem—in fact, it's one of the most visited sites in all of Iceland—but just one look at the imposing concrete structure, and it's easy to see why it's such a popular spot for worshippers and tourists alike. Completed in 1986, the church's unusual shape is inspired by the landscape and nature of Iceland—a petrified lava stalagmite, to be precise. Once you're done admiring the building itself, head up the 245-foot tower for an epic view of Reykjavík.

STEP INSIDE AN ODDLY SHAPED APARTMENT

The Cube Houses—Rotterdam, Netherlands

The first time you step out of Rotterdam's Blaak Train Station, you might feel as though you've accidentally walked into a kaleidoscope, thanks to the row of forty colorful cubic homes positioned on a forty-five-degree slant. The residential apartments, which sit above a road, were developed by Dutch architect Piet Blom in the 1970s as a creative solution to a city urban-planning issue. While the cubes might seem roomy and comfortable on paper, in reality only about a third of the floor space is usable. Curious visitors can pop into the Show-Cube museum to get an idea of what living that cube life would actually be like.

GET MARRIED IN A SHOE

High-Heel Wedding Church—Chiayi, Taiwan, China

Forget Cinderella; there's a new glass slipper in town. In a park in the city of Chiayi, Taiwan, a huge stiletto-shaped structure stretches 55 feet (18 meters) high and 36 feet (11 meters) wide and is made up of 320 panels of tinted blue glass. Despite being called a "church," no regular religious services take place within the building; the spot is used for weddings and photo shoots more than anything else. As to why the building came to be, apparently the design was intended to attract women. (No comment on that.) Regardless, it certainly is . . . something.

VISIT A DESERT CITY LIVING IN THE FUTURE

Nur-Sultan, Kazakhstan

It's too hard to pinpoint just one building here—the entire city of Nur-Sultan (formerly Astana) has become known for its incredibly futuristic architecture. Not long after it was declared the capital of Kazakhstan in 1997, the government enlisted the help of Japanese architect Kisho Kurokawa to develop new plans for the once small and remote city, which included surreal blue-and-gold buildings and wide streets. Some notable spots to visit include the 203-foot silver pyramid that houses the Palace of Peace and Accord (designed by British architect Norman Foster), the pair of golden towers (nicknamed the "Beer Cans"), the concave Shabyt building (a.k.a. the "Dog Bowl"), and the iconic Bayterek tower. Be sure to check out the skyline when it's all lit up at night; you'll feel like you've stepped into the year 3022.

UNIQUE FESTIVALS

We love music festivals; we do. But to really understand a place, its people, and its culture, there is nothing like gathering with the residents to celebrate their time-honored traditions and events beyond the merely musical. Here are some festivals we think belong on your bucket list.

WATCH AN EPIC FISHING CONTEST

Argungu International Fishing and Cultural Festival—Argungu, Nigeria

Imagine thousands of people descending into a river, splashing and scrambling to catch the biggest fish with handheld nets—and that's this festival. Aside from this epic fishing contest, the four-day celebration also offers cultural displays, agri-cultural exhibitions, music and dance performances, and swimming and wrestling competitions, among other things. It all happens once a year in the state of Kebbi in Nigeria's northwest, and aside from being extremely fun—and an important part of passing down skills through the generations—the festival is considered a symbol of maintaining friendship between Argungu and the nearby community of Sokoto.

WANDER THROUGH A WINTER WONDERLAND

Harbin Ice Festival—Harbin, China

Attention, lovers of ice and snow! This is the world's largest ice festival, a true winter wonderland of giant palaces and magical castles all made entirely with ice, some with frozen slippery slides to go down when you're done exploring their rooftops. There are massive, intricate snow and ice sculptures, and even ice lanterns, which have been a tradition in Harbin for centuries. At night, everything is lit up in colorful light displays—it's all very pretty. You can also get adventurous at the festival with skiing, sledding, fishing, and even winter swimming (if that's your thing). Brrr.

ENTER THIS CANADIAN HAIR-FREEZING CONTEST

Takhini Hot Springs—Yukon, Canada

This event gives a whole new meaning to frosted tips. You dip your hair into hot springs and then let it freeze in the icy cold air as you style it into an extravagant frosty sculpture, all while your body stays toasty in the water. Visit the Takhini Hot Springs, Yukon, during the contest (dates vary, but the contest runs for a couple of months during winter) and you can enter, provided the temperature is below -4 Fahrenheit (-20 Celsius). There are prizes for the best looks, but the real prize is maybe the wacky frozen-haired friends you meet along the way.

FILL UP ON BREAD OF THE DEAD

Festival del Pan de Muerto—Mexico City, Mexico

Around the time of Día de los Muerto (Day of the Dead) in Mexico, you'll start seeing *pan de muerto* everywhere. In its classic form it's a yeasted sweet bread, flavored with orange-blossom essence and zest and dusted with sugar. If you like it, you won't want to miss the Festival del Pan de Muerto in Mexico City, which brings together more than forty bakeries to offer dozens and dozens of variations of the delicious bread. Ferrero Rocher *pan de muerto* is a must-try, but if you're feeling especially adventurous, go for flavor combos like wine and grapefruit. It's sweet, bready heaven.

TRY YOUR HAND AT BED RACING

Sitges Carnival—Catalonia, Spain

The best type of race is surely the one where you get to lie down on a bed and be pushed along by somebody else, right? The carnival of Sitges, a small town southwest of Barcelona, is the place to make this lifelong dream happen. Sitges is known as the LGBTQ capital of Spain, and beyond the important tradition of bed racing, the carnival is quite spectacular, with lavish costumes and wild parties galore. Dance alongside the floats in the Debauchery Parade, which is exactly what it sounds like, or follow a giant sardine effigy down to the water to see it ceremoniously buried in the sand. There really is something for everyone here.

BE DAZZLED BY GOLDEN TURMERIC SKIES

Bhandara Festival—Maharashtra, India

Turmeric is known for its myriad health benefits, but in the small town of Maharashtra, it's also famous for transforming the air into gold. Every year during the incredible Bhandara Festival, thousands of devotees come to the Khandoba temple seeking the blessings of Lord Khandoba, who is worshipped with handfuls of turmeric powder and other offerings. There's *so* much turmeric tossed around that the entire town is drenched in a golden yellow hue, as hearts are purified with devotional chanting and celebration. Just be ready to leave the festival stained golden yourself. We promise it'll be worth it.

JAM OUT UNDER THE SEA

Lower Keys Underwater Music Festival—Florida Keys, Florida, United States

This might be the closest thing lovers of *The Little Mermaid* can get to a crab-conducted underwater symphony. The Underwater Music Festival in Florida's Lower Keys brings together divers and snorkelers who play pretend sea-themed musical instruments. Visitors can float through watching the performances while listening to music that is piped underwater by a local radio station. The goal of the event, aside from bringing a whole lot of underwater joy, is to raise awareness about how to preserve the Keys' fragile coral-reef ecosystems, so it's all for a good cause.

ENJOY A PENIS FESTIVAL

Kanamara Matsuri—Kanagawa, Japan

This beautifully chaotic festival involves a procession of three giant phallic shrines, and aside from the colorful parades you'll find an abundance of penis-shaped delicacies, decorations, and carved vegetables. For those who enjoy a good penis here and there, it's a fun time. But it's not just about dongs for the sake of it: the festival is a celebration of fertility and has a deeper religious meaning. Sex workers would once visit Kanamara Matsuri to pray for protection against sexually transmitted diseases. Today, profits from the event go toward HIV research, so you can feel good about all that phallic candy and other merchandise you're definitely going to leave with.

Good to Know: On the face of it a festival might seem like a whole lot of frivolous fun, but remember that it could carry a deep spiritual significance for those celebrating. Remember to ask permission before taking photos of people, and take cues from locals if you're ever unsure about how to behave.

HAVE A "BUN" TIME IN HONG KONG

Cheung Chau Bun Festival—Cheung Chau, Hong Kong, China

This lively festival happens on Cheung Chau island, a short ferry ride from the city, and is dedicated to the god Pak Tai. Bamboo towers are constructed and covered in lucky steamed buns, which are an important and symbolic local product. There's also a sixty-foot tower covered with artificial buns intended for the famous "bun-scrambling" competition, in which contestants climb up the tower trying to grab as many of the lucky buns as they can. There are also colorful parades, food offerings, gods and deities to pay respect to, and vendors selling trinkets and decorations. But let's be honest—you're mainly going for the buns.

SPRING INTO A FEAST OF FLOWERS

Medellín Flower Fair—Medellín, Colombia

Medellín is known as the "City of Eternal Spring," and when its flowers bloom in early August each year, thousands of visitors come to the city to experience folk music, dancing, and other cultural events. The one you'll really want to see is the parade of the Silleteros (Desfile de Silleteros), where farmers make giant flower arrangements attached to wooden contraptions called *silletas*, which allow them to carry their floral masterpieces on their backs. The intricate arrangements depict their lives, stories, and culture, and hundreds of flower-bearing contestants join the colorful parade as crowds cheer them on from the sidelines.

SURPRISING SPAS

Damn, soaking up all that culture is exhausting, amirite? When you're ready for a break from . . . your break, recharge with these spa experiences that are relaxing, yes, but also far from boring.

GET MASSAGED BY KNIVES

Chinese World Knife Therapy Association in Taipei City Mall—Taipei, Taiwan, China

Nowhere near as terrifying as the name suggests, knife massage, or *dao liao*, is a surprisingly relaxing Chinese massage technique that involves being repeatedly hit with a pair of blunt (not sharp!) meat cleavers. Therapists target pressure points, and along with providing physical relief, the steel knives are thought to release and absorb bad energy. It's believed to date back more than two thousand years to ancient China. These days, you can find various centers offering knife massages in Taiwan (as well as in some other countries), but this no-frills spot in an underground shopping mall is a great affordable option.

EXPLORE EUROPE'S BIGGEST SPA

Therme Bucuresti—Bucharest, Romania

Think of the biggest spa you've ever been to. Now, times it by a gazillion. That's the kind of scale we're talking about at Therme Bucuresti, one of the biggest spas in all of Europe. Located about ten minutes from the city of Bucharest, this indoor tropical paradise is a balmy 84 to 86 degrees Fahrenheit (28 to 30 degrees Celsius) year-round and boasts ten pools, ten saunas (six dry and four wet), as well as a whopping sixteen waterslides, which all use thermal water sourced from deep underground. The massive complex is also home to Romania's largest botanical garden, which contains more than eight thousand plants.

Travel Tip: If you can't make it to Romania, look up some of the other Therme spa locations around the world—you might be able to find one closer to home.

EAT SWISS FONDUE IN A HOT TUB

Alpenbad—Hinwil, Switzerland

Arguably the only thing better than soaking in a hot tub with a view of the snowcapped Swiss Alps is doing so while simultaneously stuffing your face with cheese. You can experience all of the above at Alpenbad, a family-run alpine bath located about an hour outside of Zurich. The tubs are filled with mineral-rich Bachtel springwater (not cheese; sorry), sourced from the property's on-site spring and heated to a toasty 102 degrees Fahrenheit (39 degrees Celsius). You can order wine and various regional dishes directly to your private hot tub, but let's not forget why we're all here: the fondue. This house specialty is made from a mix of five different cheeses from a local cheese maker and served with two types of bread. Oh, and it's lactose-free!

SOAK IN A BATH OF BEER

Thermal Beer Spa—Budapest, Hungary

Budapest is known for its spas and bathhouses, but this one has a boozy twist. Warm water (about 97 degrees Fahrenheit, or 36 degrees Celsius) is combined with natural beer extracts—namely, hops, malt, and yeast—and the minerals, vitamins, and oils within this mixture are thought to provide skin and body benefits, such as reducing the appearance of acne, cleansing pores, and releasing muscle tension. As you soak in the wooden tubs, you can pour yourself unlimited (!!) craft ales from your own personal beer tap. There are two locations in the city to choose from: Lukács Bath on the Buda side and Széchenyi Bath on the Pest.

SPEND TWENTY-FOUR HOURS IN A KOREAN SPA

Wi Spa—Los Angeles, California, United States

If you take your "spa days" literally, you'll be pleased to know you can spend an entire twenty-four hours at this Korean-style spa in Los Angeles. This massive complex offers classic treatments like massages, facials, and nail and skin care, but the real drawcard is the themed sauna rooms. From the clay sauna, where guests can lie on thousands of tiny clay balls that are believed to help with detoxification, to the salt sauna that's supposed to benefit the respiratory, circulation, immune systems, to the ice sauna that brings your temperature back down and is thought to tighten pores and improve overall well-being, you'll have no trouble spending your whole day relaxing here; trust us.

HEAL YOUR AILMENTS IN A SALT CAVE

Asheville Salt Cave—Asheville, North Carolina, United States

Ever heard of salt therapy? In the small mountain town of Asheville, North Carolina, you can relax, meditate, or get a massage in a cave filled with twenty tons of pure, pink salt. The temperature and humidity of the cave are carefully monitored to replicate the microclimate of a salt mine. Breathing in the salty, negative ion–rich air is thought to provide an array of healing benefits, such as an immune system boost or relief from respiratory ailments like asthma. Plus, the blush-pink crystals that line the walls look truly magical.

DISCOVER A QUIRKY SPA THEME PARK

Yunessun—Kanagawa, Japan

Up in the hills overlooking the Hakone mountains is Yunessun, a hot-springs spa meets theme park that you could only find in Japan. There are twenty-three different baths ranging from traditional-ish no-clothing-allowed Japanese *onsens* to kitschy themed pools, including ones containing sake, coffee, or even wine. If your feet could use some attention, pay a visit to Dr. Fish Foot Bath for an exfoliation treatment courtesy of hundreds of hungry little fish that'll nibble away at your dead skin (which is kinda gross but also strangely satisfying?). Once you're done relaxing, you can get your heart rate back up with a few rides on the water slides or a swim in the pool.

Psst! For more theme parks, flip to page 124.

GO ON A THERMAL SPA CIRCUIT

Piedra de Agua—Baños, Ecuador

Why have one spa treatment when you could have eight? In the aptly named town of Baños (meaning "bathing"), Piedra de Agua offers a unique spa circuit that allows guests to soak up the natural benefits of thermal water, steam, and volcanic mud. Start by slathering yourself with volcanic thermal mud in the red mud pool (thought to rejuvenate the skin) before dipping into the blue mud pool, which contains small quartz crystals (thought to help with energy alignment). After, you can venture to the underground cave pools, plonk yourself into a wooden steam box, and check out the Turkish- and Japanese-style baths. Once you're all spa-ed out (if that's even possible?), relax in one of the outdoor pools and have cocktails delivered to you on a cute lil' floating drink boat.

TAKE A HAY BATH (YES, IT'S A THING)

Hotel Heubad—Fiè allo Sciliar, Italy

If you're prone to seasonal allergies, feel free to skip reading this one and jump ahead to the next. Okay, still with us? Here it goes: Hotel Heubad, a resort tucked away in the mountains of northern Italy, specializes in "hay baths"—spa treatments that involve being wrapped up with warm, damp hay sourced from the Dolomites. The heat is thought to open pores, allowing the hay to penetrate the skin. The treatments date back more than a century, when visitors used to rest their weary heads on beds of hay and wake up feeling relaxed and energized. If just the thought of being covered in piles of hay is enough to make your skin itchy, don't worry—there are other, less grassy treatments on offer, too.

COLORFUL PLACES

What's life without color? From entire towns drenched in a single hue to Mother Nature's curiously colorful secrets, there are rainbow mysteries in every corner of the globe that are waiting to be discovered.

CLIMB A RAINBOW MOUNTAIN

Vinicunca—Vinicunca, Peru

Rainbow Mountain, known as Vinicunca in the local language, is a striking mountain in the Peruvian Andes that's covered in a magnificent array of colors. The mountain gets its pigment naturally from mineral sediments in the area and seems to be cloaked in a striped blanket of reds and golds, lavender and turquoise. At more than sixteen thousand feet (~4876 meters) above sea level, it's a decent trek, but if at any point it becomes too much you can request a horse to help you to the top. Plus, there are a lot of cute alpacas and llamas to spot along the way. The mountain may be a product of climate change—it was apparently covered in snow until recently, when the snow melted away to reveal its true colorful self.

Good to Know: Speaking of climate change, global tourism has a hefty carbon footprint. You can learn more about traveling responsibly and sign a "Travel Better" pledge at sustainabletravel.org. For more eco-conscious travel ideas, flip to page 58.

SEE THE BRIGHTEST PART OF THE ARCTIC

Kangaamiut—Qeqqata, Greenland

Despite its name, Greenland may not be the first place that comes to mind when you think of colorful destinations—but on the rugged western coastline at the mouth of the Eternity Fjord, a rainbow surprise awaits. Kangaamiut is a small village accessible by boat and a favorite for heli-skiers. And as with many other villages in Greenland, brightly colored wooden houses dot the mountainous coastline, a fairy-tale sight at all times but especially when the terrain is covered in snow. It's a picturesque place to pass by for a photo or two, but an even better place to stop awhile to get to know the community.

VISIT A VIBRANT ISLAND VILLAGE

Nubian Village—Aswan, Egypt

This small village on Elephantine Island—along the Nile River near the city of Aswan—holds a long and fascinating history. Nubian Village, which also contains the palm grove–filled Siou and Koti villages, is home to one of the remaining Nubian communities in southern Egypt and Sudan. The village is best explored on foot, and wandering the garden paths you'll find richly decorated colorful dwellings painted with geometric shapes, crocodiles, and other symbols of Nubian culture and local families. You might spot a real crocodile or two, and you'll definitely see camels. It's a vibrant and mystical place to spend the day.

FLOAT DOWN A COLORFUL CANAL

Xochimilco—Mexico City, Mexico

South of Mexico City is a magical place called Xochimilco, where colorful boats make their way down canals past traditional Aztec *chinampas*, or floating gardens. The gondola-like boats, called *trajineras*, are painted in vibrant colors with big floral arches and can seat about twenty people. Sometimes a boat will drift by with a mariachi band aboard, and if you tip them enough, they'll come over and play a few extra songs for you. It's a popular place for big groups to hang out, so if you're looking for a peaceful canal ride, it's better to visit during the week and earlier in the day. Otherwise, bring beers and be ready for a fun time!

GET SERENE IN A BLUE MOUNTAIN VILLAGE

Chefchaouen, Morocco

Ready for a whole lot of blue? Chefchaouen is the famous blue village in Morocco, nestled into the side of the Rif Mountains. Nearly every wall and surface is bathed in a glorious shade of blue, which was originally attributed to the city's Jewish residents. Wander the narrow alleys of the old medina to arrive at the Kasbah with its lush garden. Sip on mint tea and sample the tagines, kefta, and freshly baked breads. The walking paths up the mountain give an even better view of this blue beauty, along with plenty of cactus and wildflowers. Hello, serenity.

JUST THREE PINK LAKES

One of Mother Nature's prettiest gifts is surely the assortment of candy-pink lakes around the world. Why are the lakes pink, you ask? It's usually because of their extremely high salt content, which means the only organism that will survive in their waters is a type of micro-algae that turns the water a rosy hue. The lakes aren't always pink year-round, so make sure to do some research before you go.

1. LAKE RETBA—DAKAR, SENEGAL

Also known as Lac Rose, this pink watery surprise is about an hour's drive from Senegal's capital, Dakar. Along its banks you'll see piles of salt, which locals extract from the lake.

2. LAKE HILLIER—MIDDLE ISLAND, AUSTRALIA

This pink beauty is right next to blue ocean, white sand, and a strip of green forest—which makes it an amazing sight to take in from a scenic flight.

3. LAS COLORADAS—YUCATÁN PENINSULA, MEXICO

A striking pink lagoon that's also the result of salt production in the area, Las Coloradas attracts flocks of flamingos, who show up to get some of that sweet, sweet algae.

Fun Fact: *The reason flamingos are pink is because of the algae and shrimp—found in lakes and lagoons like Las Coloradas—that they munch on all day.*

VISIT CHINA'S CRIMSON WETLANDS

Panjin Red Beach—Panjin, China

If you're a fall foliage lover, here's something different for you. Red Beach is a giant wetland ecosystem in the Liaohe Delta, near Panjin City in Liaoning Province, China. The seepweeds that grow on the wetland's mudflat turn a striking shade of crimson in the fall, a romantic red carpet covering the marshy landscape. The wetlands are home to a huge variety of bird species and other wildlife, and nearby golden rice paddies can be seen from the Red Beach Wharf. Oh, and—plot twist—there's no actual beach here, so you can leave your bathing suit at home.

Travel Tip: The Red Beach is beautiful year-round, but be sure to visit in the fall if you want to see the iconic crimson hue. In the mornings you'll get the best photos.

EXPLORE ISTANBUL'S COLORFUL GEMS

Fener & Balat—Istanbul, Turkey

Balat is the traditional Jewish quarter in Istanbul's Fatih district. Here you can wander narrow cobbled streets lined with tall wooden houses in a rainbow of colors. Brightly painted gourds hang from terrace balconies, and around every corner is a hidden pocket of color and art. There are cute, artsy cafés and restaurants to explore, and it's right next to the former Greek neighborhood of Fener, which is known as the Vatican of the Greek Orthodoxy. The colorful area between Yildirim and Vodina Streets is a good place to start here.

GAZE AT A MYSTERIOUS GEYSER

Fly Geyser—Gerlach, Nevada, United States

At the edge of Black Rock Desert, the earth seems to have risen into a surreal, multicolored tower that spits steam and hot water down its sides. This otherworldly sight is a geothermal geyser—it was originally man-made, but nature has taken over now—and it's located on the privately owned Historic Fly Ranch in Nevada. As the Fly Geyser spews water, it deposits minerals on its calcium carbonate surface, which invites the growth of tiny colorful algae, and the result is an astonishing palette of reds, yellows, and greens. You'd be forgiven for thinking you're on another planet when you spend time in the presence of this thing.

VISIT COLOMBIA'S MOST COLORFUL TOWN

Guatapé, Colombia

Guatapé is a unique town near Medellín, famous for its brightly colored houses with painted *zocalos*. The buildings have elaborate frescoes painted along their lower facades, which depict animals, flowers, family stories, and other scenes of life in the region. And speaking of otherworldly geological formations: don't miss the giant Piedra del Peñol, also known as the Rock of Guatapé. It's an enormous (like, really big) rock you can climb that gives spectacular views of the nearby towns, bright-blue lakes, and lush green islets. From the frescoes to the flowerpots to the views, everything is drenched in color here.

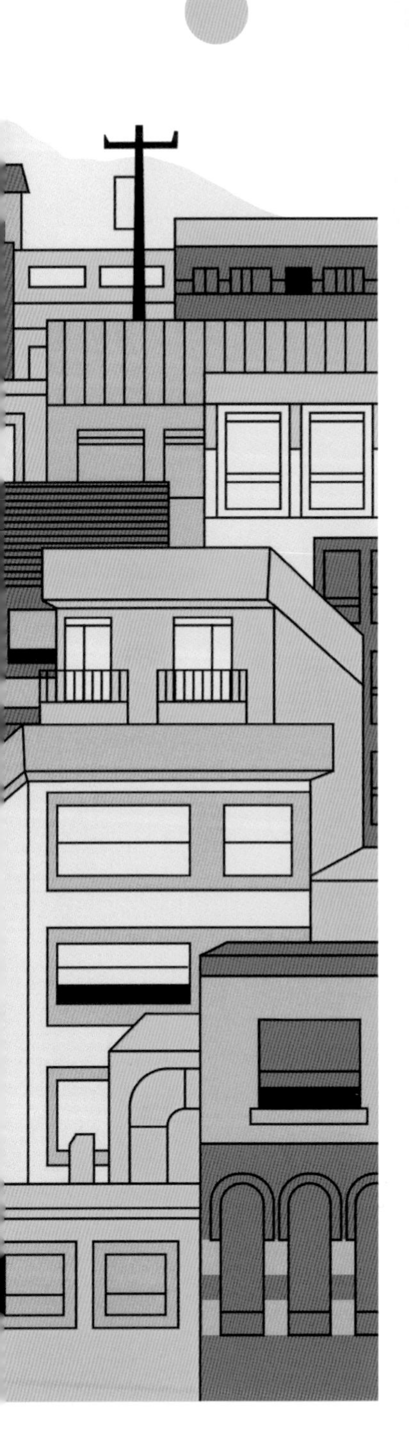

Good to Know: There's been a trend in recent years of local governments revitalizing their towns by allowing artists and residents to give every building a colorful paint job. Gamcheon Culture Village in Busan, South Korea, and Indonesia's Kampung Pelangi "Rainbow Village" are two examples of villages transformed into multicolored dreamlands. So if you need more color in your life, add these to your bucket list.

TAKE A DIP IN A TURQUOISE OCEAN TRENCH

To Sua—Upolu Island, Samoa

To Sua is an unusual circular ocean trench and natural swimming pool found on the property of a local couple on the southeast coast of Samoa. It's a vivid turquoise beauty with crystal-clear waters, surrounded by lush foliage and tropical gardens. Dive in and swim alongside tropical fish, and when you're done splashing around, climb back out to enjoy the sweeping views saturated in Mother Nature's best greens and blues. There are a rock pool for kids, fishing spots, and a great beach nearby, too.

WALK DOWN LISBON'S FAMOUS PINK STREET

Rua Nova do Carvalho—Lisbon, Portugal

You probably won't find any more brothels in Lisbon's Cais do Sodré—what was once its red-light district has now been transformed into a lively stretch of cafés, bars, and restaurants all along a bubblegum-pink street called Rua Nova do Carvalho. Get there before sundown if you want a photo on the iconic pink walkway, but stay till after dark if you want to enjoy the buzzing nightlife that the neighborhood is known for. Lisbon seems to love pink, and once you realize this, you'll start seeing pink foods and drinks everywhere. Pink lattes, anyone?

WEIRD MUSEUMS

Sure, the Tate and the Met are great and all, but how about an entire museum dedicated to toilets? These wonderfully weird collections dive into sometimes strangely specific topics and are bound to make you say, "Huh, interesting!" numerous times—we guarantee it.

SEE MATH IN A NEW WAY

National Museum of Mathematics—New York, New York, United States

"Fun" and "math" might not be words you associate closely with each another, but that may change after a trip to New York City's Museum of Mathematics. The two floors are filled with brain teasers, fascinating puzzles, and interactive exhibits that use math to make seemingly impossible things—like riding a tricycle with square wheels or sitting on a chair that never tips over—actually possible.

STEP INTO AN OTHERWORLDLY LIGHT SHOW

MORI Building Digital Art Museum—Tokyo, Japan

This digital art museum takes "interactive art" to a whole new level. Vibrant, futuristic light displays are projected on the walls, floors, and ceilings of a series of connected exhibit rooms. People can walk freely throughout the space and not only interact with the artworks, but also watch as the works interact with each other. The result is a truly mind-blowing experience that'll have you feeling like you're living in an episode of *Black Mirror*.

EXPERIENCE THE WONDERFUL WORLD OF LAWNMOWERS

British Lawnmower Museum—Southport, England

Honestly, it doesn't get more British than this museum dedicated to a seemingly mundane piece of gardening machinery. After you've learned all about the (surprisingly interesting) history and development of the lawnmower, you can check out the extensive collection on display—including the *Lawnmowers of the Rich and Famous* exhibition featuring one rather regal mower owned by Prince Charles and Princess Diana.

PERUSE PLENTY OF PENISES

The Icelandic Phallological Museum—Reykjavík, Iceland

Yes, this is exactly what it sounds like. This penis museum boasts a sizable collection of more than two hundred penises and penile parts, sourced mostly from local Icelandic land and sea mammals (think specimens from polar bears, seals, walrus, and sixteen different species of whale, plus displays from over twenty types of land mammals—*Homo sapiens* included). But it's not all biology: there are also tons of artworks, objects, and utensils, all in keeping with the museum's phallic theme, of course. When you're done browsing the exhibits, visit the gift shop to pick up souvenirs for all your loved ones back home. (You're welcome, Grandma.)

SATISFY YOUR NOODLE CRAVINGS

CUPNOODLES Museum—Osaka, Japan

Learn all about what is, without a doubt, the single most important culinary invention of our time: instant noodles in a cup. Check out the history exhibit, the cup noodle–shaped theater, and the instant-noodle tunnel filled with around eight hundred different types of packaging. And since you'll no doubt be in the mood for noods after browsing the museum, make sure you stop by the factory to create your own ~original~ cup noodles, customizing everything from the broth and the toppings to the cup itself.

EMBRACE YOUR INNER CAT LOVER

KattenKabinet—Amsterdam, Netherlands

If you spend 90 percent of your time on vacation missing your cat at home, this one's for you. At KattenKabinet—which literally means "cat cabinet" in English—you can pay homage to your fave feline friend and browse a collection of artworks that explore the role of cats in art and culture. The museum, which sits on the first floor of a historic canal house from the 1600s, consists of several differently styled rooms filled with centuries-old paintings, sculptures, posters, drawings, and lithographs, all with one common theme: the cat.

BROWSE A BIG COLLECTION OF SMALL BOOKS

Museum of Miniature Books—Baku, Azerbaijan

There are no prizes for guessing what the main attraction is here: teeny-tiny books. What started as a private collection of Zarifa Salahova, a self-confessed bibliophile, is now the only museum of its kind on Earth—and it's free to the public. The building itself is just a couple of rooms (they are mini books, after all), but the collection is massive, consisting of more than six thousand miniature publications sourced from all across the world, including the three smallest books known to exist. Better pack your magnifying glass for this one!

PEEK INTO BREAKUPS AROUND THE WORLD

Museum of Broken Relationships—Zagreb, Croatia

Ever held on to a box of mementos after a breakup? This museum is kind of like that—on steroids. While it might sound a little strange—an exhibition filled with random things that remind people of their exes—and possibly a bit sad, it's actually quite a lovely concept. All the seemingly mundane, everyday objects are really just symbols that give viewers a peek into (defunct) relationships the world over. Some stories are beautiful or funny, others are shocking or tragic, but they all have one thing in common: they're real explorations of human connection and the complicated thing called love. Who knew an old ticket stub could carry so much baggage? (P.S. There's also an outpost in Los Angeles.)

PONDER THE MEANING OF DEATH

Museum of Death—New Orleans, Louisiana, and Los Angeles, California, United States

Speaking of things you need to do before you die, if you can muster the courage, you should definitely pay a visit to this morbid yet fascinating museum. The original location was opened in San Diego's first mortuary in 1955 to educate and encourage more open conversations about death. The museum now has locations in NOLA and LA, and both spots feature unique collections, including letters and artworks created by serial killers, real crime-scene photographs, tools used by morticians and coroners, and skulls galore.

LEARN THE HISTORY OF LOOS

Sulabh Museum of Toilets—New Delhi, India

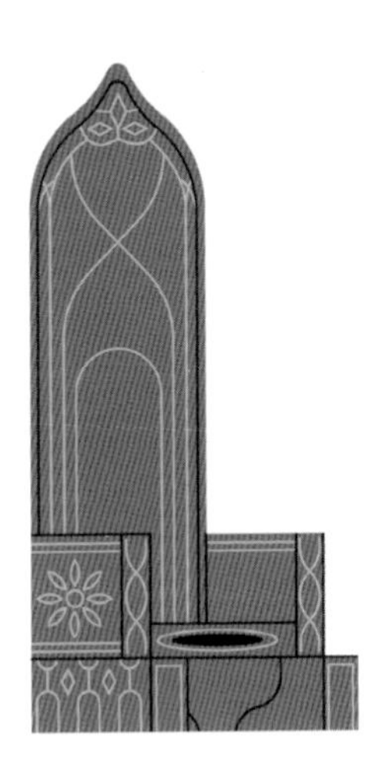

Don't write this one off just based on the name. We promise no poop jokes here—they stink. On the surface it may seem like just another odd-sounding, quirky museum, but it's tied to a greater purpose: to help improve India's sanitation conditions through education. Owned by Sulabh International Social Service Organisation, a nongovernmental organization dedicated to making sanitation and water available to all, the museum follows the interesting history of the toilet through ancient, medieval, and modern times. Highlights include an intricately carved urinal, a replica of a treasure chest–shaped loo, and a plethora of fun facts.

DISCOVER INTERESTING HUMAN ODDITIES

Kunstkamera—St. Petersburg, Russia

Founded by Peter I (yes, *that* Peter) more than three hundred years ago as part of an effort to modernize Russia, Kunstkamera, the country's first public museum, is well-known for its huge collection of human oddities. While many people nowadays might describe the display of obscure anatomical specimens—like preserved human embryos, skeletons of conjoined twins, and more than a thousand preserved animals—as insensitive, the museum was reportedly never intended to be a place for entertainment but rather an important scientific documentation of human development. Either way, you'll remember a trip here for years to come.

QUIRKY TOWNS

Sometimes you've just gotta get out of the big cities and enjoy the slower pace of small-town life. But not every town is created equally: the world has been blessed with countless cute, quirky, charming, and strange towns and villages. Here are some of our favorites.

GET COZY IN A BOOK TOWN

Hay-on-Wye, Wales

Bibliophiles, gather 'round. Hay-on-Wye is a book town—yes, this is a thing—and is famous for its many antiquarian and secondhand bookstores. In the summer it attracts big crowds for the literary Hay Festival (Bill Clinton once called it the "Woodstock of the mind"), but all year long you'll find cozy bookshops, eateries in which to read purchased books, and even open-air honor-system bookshelves to peruse. There's a bookstore dedicated entirely to horror and mystery stories and another just for poetry. And for those looking for more than just a good read, this quaint Welsh town has some great antique stores to explore as well.

SEE THE WORLD'S MERRIEST CEMETERY

Săpânța, Romania

If visiting a small town for a cemetery alone seems odd to you, you probably haven't been to Săpânța. This small village in Maramureş, one of the more traditional regions in Romania, is famous for the way it joyfully honors its dead. Behind the Church of Assumption is what's known as the Merry Cemetery, a striking graveyard with more than seven hundred carved wooden grave markers painted in lively, colorful designs that honor the lives and foibles of those resting peacefully below. Each richly decorated cross also features a poem: some witty and humorous, some heartbreaking, and all brutally honest. In life and in death, few secrets are safe in this small town.

SLEEP IN AN ITALIAN CONE HOUSE

Alberobello, Italy

This small Italian town is a UNESCO World Heritage site known for its abundance of tiny *trulli*—circular limestone dwellings with curious cone-shaped roofs. The structures, which can date from as early as the fourteenth century, are characteristic of the Valle d'Itria region in Puglia, southern Italy. You'll wander narrow streets lined with charming stores and pause in trattorias with alfresco dining to sample delicious local foods. Alberobello is a fairy-tale town and one of many other quaint villages to visit in the region. With its rolling countryside and proximity to stunning beaches, Valle d'Itria is definitely worth spending some time in.

STOP BY EARTH'S NORTHERNMOST COMMUNITY

Longyearbyen, Norway

This tiny Norwegian town is unlike anywhere you've been before. Surrounded by Arctic wilderness, it's a starting point for nature adventures in Svalbard and a place where reindeer wander the streets, whales swim by in the fjord, and polar bears are never far away (seriously, the locals carry weapons whenever they leave the settlement just in case a polar bear makes itself known). People don't spend their entire lives in Longyearbyen—both giving birth and being buried here are forbidden for safety reasons—but those residents who do stay awhile are warm and hospitable to anyone passing through.

TAKE A FERRY TO CAT ISLAND

Aoshima, Ehime, Japan

Cat people, this is for you. Aoshima is a tiny island in Ehime Prefecture, Japan, known as Cat Island due to its more than 120 feline residents who rule over only a handful of human subjects. The cats own the place, occupying abandoned buildings, lazing in the sunshine, and occasionally enjoying a spot of fishing. They'll wander down to the ferry wharf in large numbers when you arrive—to greet you and assess your worthiness for entering their kingdom. Please don't feed the cats—this is taken care of by the elders who live on the island. But cat toys and selfies are more than welcome.

Good to Know: There are two Aoshima Islands in Japan. The other one is outside Miyazaki City and is known for its white beaches and subtropical jungle.

DISCOVER AN UNDERGROUND CITY DOWN UNDER

Coober Pedy, Australia

This fascinating town in the Australian outback is known as the opal capital of the world due to its history as an opal-mining community. But its real draw is its status as an "underground city"; an estimated 50 percent of Coober Pedy locals live underground in "dugout" homes in the hillsides, which keep them cool in the desert heat. Its ruddy, moon-like landscape made it the perfect setting for *Mad Max Beyond Thunderdome*, which was filmed there in the '80s. And because this is Australia, you should expect to find kangaroos, dingoes, and other intriguing wildlife in the area.

WANDER A VILLAGE IN THE ROCKS

Setenil de las Bodegas, Spain

Andalusia is famous for its many white villages, or *pueblos blancos*, but this hidden town takes the cake. Setenil de las Bodegas was built into a network of caves in a cliff, dwellings that have long been favored for their natural coolness and used as bodegas to store food and wine. The village's whitewashed buildings seamlessly blend into the rocky landscape, and almond and olive trees spring up from the rooftops and surrounding mountains. Today you can still find tucked-away bodegas that offer local food products and wine from the region and an abundance of charming restaurants and bars to stop in, pre- and post-siesta.

ADMIRE THESE TRADITIONAL PAINTED MUD HOUSES

Tiébélé, Burkina Faso

This circular village is known for its *sukhala*, curved, windowless houses made of mud and clay with grass-thatched roofs. What's most striking about the dwellings and their connecting walls is that they're all hand-painted with intricate designs and symbols that represent the spiritual and cultural identity of the region. The town is inhabited by the Kassena people, one of the oldest ethnic groups located in southern Burkina Faso along the Ghanian border. There are a few simple guesthouses in the village, which is accessible by bus from the cities of Pô and Ouagadougou.

VISIT A VILLAGE WHERE PLANTS REIGN

Houtouwan, China

China is known for its emergence of eco-cities, but only ghosts live in Houtouwan now. It was once a fishing village but, being so isolated, was slowly abandoned in the 1990s. No problem, though; this allowed plants to naturally reclaim the village, and now every building is covered in a blanket of ivy and other lush green vegetation. It's an eerie place, with buildings seemingly sprouting from the hillside like plants themselves. But it's visually spectacular, especially if you have a fondness for green. Houtouwan is located about forty miles from Shanghai on Shengshan Island in China's eastern Zhejiang Province.

PARTY DOWN IN SOUP TOWN

Supilinn, Estonia

That's right, in Estonia there's a place called Soup Town. Supilinn is the historic district of southern Estonian university town Tartu. It's a former slum town, established in the eighteenth century, and it survived a lot of the World War II destruction that the rest of the city did not. The streets are all named after fruits and veggies (Pea Street and Potato Street are two favorites), and the old wooden houses are a fascinating glimpse into the past. For a town named after soup, there is a surprising lack of eateries dedicated to warm bowls of liquidy comfort, but Tartu has a lot of highly rated restaurants where you can get your fix.

GIANT THINGS

You know what they say: bigger is better! The world has no shortage of ginormous man-made structures. Some have a deeper meaning, and some seem to serve no purpose other than providing a fun photo backdrop—we're looking at you, giant golf tee—but they're all impressive in their own way.

SEE THE TEMPLE OF THE GIANT SWING

Bangkok, Thailand

Wat Suthat, one of the most beautiful and revered Buddhist temples in Thailand, is also home to a giant red swing that stands proud at the entrance. It is made from teak and is an impressive sixty-five feet (twenty meters) tall. Legend has it that it was once part of a dangerous ceremony during which people would try to swing high enough to grab a sack of gold attached to a bamboo pole. Now only the red frame remains, but it's an interesting sight nonetheless.

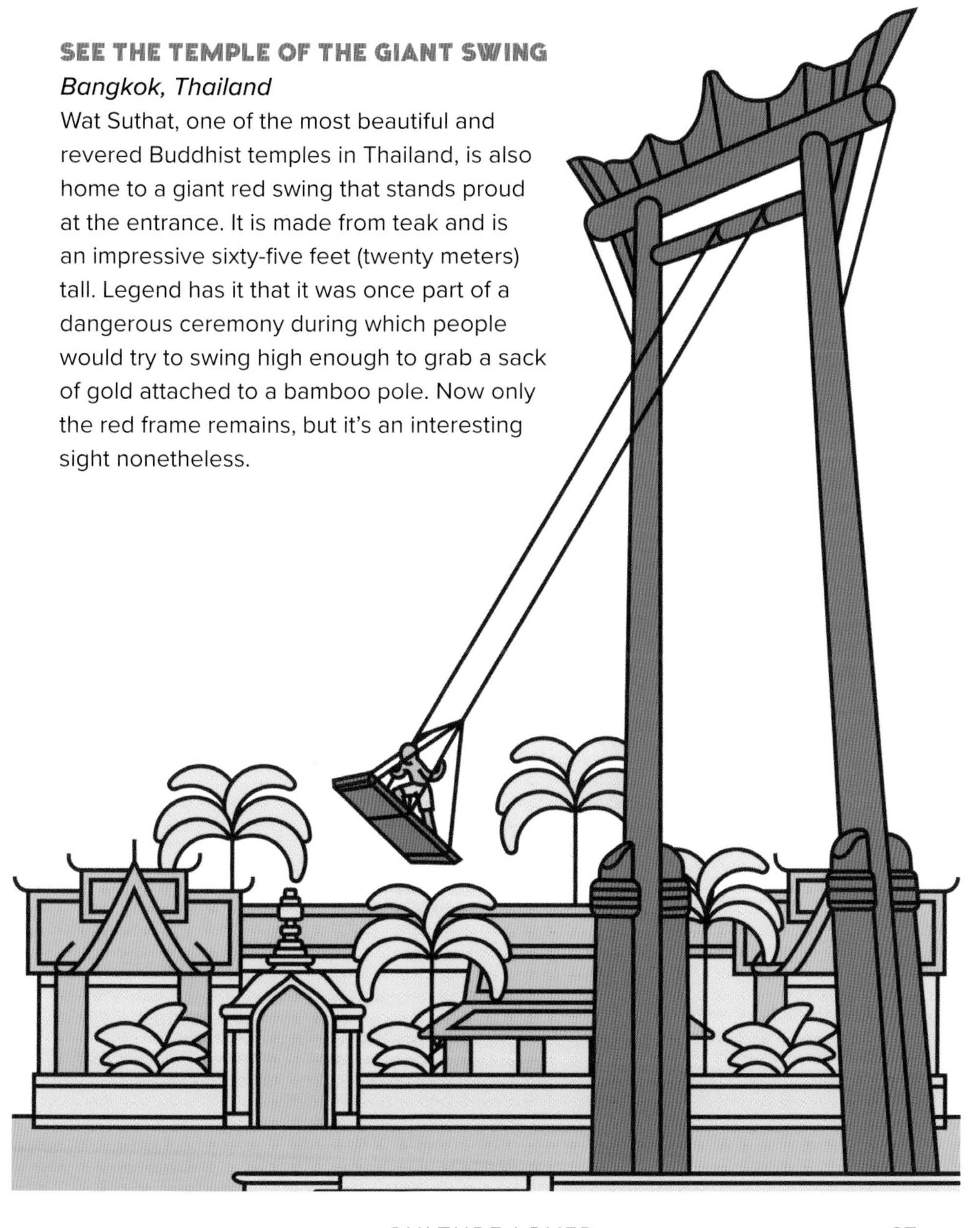

GO ON A HUNT TO FIND THE SIX FORGOTTEN GIANTS

Copenhagen, Denmark

You want *giant*? Look no further. In the western suburbs of Copenhagen, you'll find no fewer than six literal giants hiding away in the woods. Created by Danish artist Thomas Dambo (with help from local volunteers), each sculpture uses wood scraps scavenged from old pallets, fences, and sheds and demonstrates the power of recycling. To find each of the sites you'll need to follow a treasure map, and take note of the poems found near each giant; these contain clues that'll help guide you along the trail. Not only do the sculptures look cool, but they're also a fun way to explore the less obvious parts of Copenhagen.

TICK OFF A LOT OF BIG THINGS IN A SMALL TOWN

Casey, Illinois, United States

If you only have limited time in your travel itinerary for visiting extremely large, and, let's be honest, kind of random roadside attractions, this place is gonna give you the best bang for your buck. In a single pit stop, you can knock out a total of twelve of the world's largest things, including the world's largest wind chime, golf tee, mailbox, rocking chair, and barbershop pole. And if you're still after more, you'll find a number of additional big things in this tiny town, from crochet hooks and ears of corn to a mouse trap and horseshoe.

SIP WINE ON AFRICA'S BIGGEST RED CHAIR

Rooiberg Winery—Robertson, South Africa

If "sit on the largest red chair in all of Africa while enjoying a glass of pinotage" wasn't already on your bucket list, we're (a) shocked and (b) here to inform you that you should add it, pronto. You'll find this very big seat at Rooiberg Winery in Robertson, South Africa, so climb up (with your wine, naturally), snap a pic, and enjoy the view.

TELL TIME USING THE WORLD'S LARGEST SUNDIAL

Jantar Mantar—Jaipur, India

Jantar Mantar, a centuries-old astronomical observatory in the north of India, is home to a massive stone sundial called Samrat Yantra. Built in the early 1700s, this "Supreme Instrument" is important not just because of its grand scale (it's more than seventy feet high!) but because when it was constructed it was able to measure time more precisely (with an accuracy of two seconds) than any other known piece of equipment.

ADMIRE DUBAI THROUGH A GIANT PICTURE FRAME

Dubai, United Arab Emirates

This impressive golden structure is going to re*frame* (sorry) the way you see Dubai. The Dubai Frame has two towers, which are almost 500 feet tall, joined by a 328-foot bridge, and features an opaque glass viewing bridge that becomes clear when you walk over it. Touted as the world's largest picture frame, it connects the city's past and present; look to the north to see the old historic district, or turn south for views of new Dubai's ultramodern skyline. Unfortunately, there's also controversy over the origin of the design: Mexican architect Fernando Donis won a competition for his proposal of a similar frame structure, but claims he was later cut out of developing the project.

Photo Tip: Let's be real—the main reason you're going to want to visit these spots is to snap a photo. So, to make sure your pic is worth the trip, try playing around with scale. Including a regular-size object or human in your photograph can help accentuate just how giant that giant thing is!

ROAD TRIP THROUGH AUSTRALIA'S BIG THINGS

Australia has so many oversize versions of everyday objects, it's too hard to pick just one, so we're going to take the easy option and tell you about a few of them.

1. **The Big Banana in Coffs Harbour:** Located on the north coast of New South Wales, this might be the most iconic of Australia's giant things. It's everything you'd imagine it to be—big, yellow, and banana-y.

2. **The Big Mango in Bowen:** It weighs 7 tons and measures 26 feet (8 meters) wide and 33 feet (10 meters) tall and cost $90,000 to build.

3. **The Giant Koala in Dadswells Bridge:** It's 46 feet (14 meters) tall, weighs around 12 tons, and features moderately creepy red eyes.

4. **The Big Prawn in West Ballina:** Thirty-nine tons of prawn? Now that's one jumbo crustacean. Whatever you do, just don't call it a shrimp.

5. **The Big Merino in Goulburn:** Or "Rambo," as it's fondly known by locals. At 50 feet (15.2 meters) high and 59 feet (18 meters) long, this 107-ton structure is a monument to the district's wool industry.

6. **The Big Boxing Crocodile in Humpty Doo**: This giant 42-foot (13-meter) croc reportedly came with an even bigger price tag—$137,000!

7. **The Big Golden Gumboot in Tully:** Awarded to the Wettest Town in Australia, this big boot stands proudly at 26 feet (7.9 meters) tall, a tribute to the town's annual rainfall record of 7,900 millimeters reached back in 1950.

8. **The Big Galah in Kimba:** The 24-foot (8-meter) rose-chested cockatoo is perched at the halfway point between the west and east coasts of Australia.

PART TWO

NATURE WANDERER

The planet is rich in natural treasures: breathtaking forests, curious animals, otherworldly geological formations—and that's just what's on land. We're not drawn to nature travel just for the photo opportunities; science tells us that immersion in nature seriously improves our well-being, too.

Yet we can hardly encourage travelers to get out and discover Earth's natural beauty without acknowledging how fragile and endangered our natural planet really is. In 2019 a UN report warned that around one million plant and animal species are at risk of extinction—from overfishing, pollution, habitat destruction, and climate change.

So yes, by all means, go out and explore. Breathe in the cool air of a rain forest, gaze up in wonder at the Milky Way in a Dark Sky Park, dive into clear waters and meet the residents of the underwater world. But every time you do this, please remember your impact. Do everything you can to make sure that future generations will be able to have the amazing experiences in the natural world that we are still able to enjoy. We must protect and cherish our natural world while we still can.

PLANT PARADISES

Ancient trees, bamboo forests, and carpets of wildflowers that extend to the horizon—we don't deserve the botanical gifts that exist in this world. We've rounded up some of our favorite gardens, forests, and other leafy locations that should be on every plant lover's bucket list.

DISCOVER A BLUE GARDEN OASIS IN MOROCCO

The Jardin Majorelle—Marrakech, Morocco

There's nothing we love more than a garden hidden behind walls, and this colorful garden in Marrakech is one of the most beautiful you'll find. The Jardin Majorelle was designed—and then populated with botanic specimens over the span of forty years—by French painter Jacques Majorelle in the 1920s, and it was later purchased and restored by both fashion designers behind Yves Saint Laurent. Its walls are painted in ultramarine blue with brightly colored trimmings, and the garden features several hundred plant species from five continents, with cactus, tropical flowers, banana trees, palms, bamboo, aquatic plants, and more all covering the landscaped oasis.

STAND BENEATH MASSIVE TREES ON THIS ISLAND IN CANADA

Vancouver Island, British Columbia, Canada

Vancouver Island is a lush, green island just off the coast of Canada, about sixty miles west of Vancouver. It's an awe-inspiring place, with trees covered in puffy green moss, secret waterfalls, beautiful beaches, and wildlife galore. The island's temperate climate and heavy rainfall mean it's the perfect place to find dense rain forests and groves of giant old-growth trees. Visit Cathedral Grove to find trees about eight hundred years old; the largest is 182 feet tall, with a girth of about 60 feet. So majestic!

Where to Stay: There's an abundance of good places to stay on Vancouver Island, including cottages, resorts, lodges, and beachside camping. We love Wya Point Resort on Ucluelet First Nation's traditional territory, where you can sleep in a yurt next to a private beach surrounded by old-growth forest. Want to sleep in a spherical tree house? See page xvi for another unique accommodation option on Vancouver Island.

TOUR THIS CACTUS DREAMLAND IN SOUTHERN MEXICO

Jardín Etnobotánico—Oaxaca, Mexico

If you're fond of a good cactus—and who isn't?—mark this place as a must-visit. The Jardín Etnobotánico (Ethnobotanical Garden) in Oaxaca City features hundreds of species of cactus, all indigenous to the Oaxaca region. You need tickets to enter, and a guide will walk you through in a group and explain the relationship between the plants, people, and cultural traditions of Oaxaca. At one point you'll probably want to linger for photos alongside the incredible rows of tall castus that form a corridor. But pay attention to the smaller fries, too; you'll find so many fascinating (and photogenic) plants in this garden.

Good to Know: You can only enter the Jardín Etnobotánico at a set time and with a guide. Tours are offered in Spanish or English. Visit the garden's website, jardinoaxaca.mx, for the schedule.

ARIZONA, UNITED STATES

These Flowers in the Sonoran Desert Bloom Only One Night a Year, and Nobody Really Knows Why

The Night-Blooming Cereus (or Queen of the Night) is a species of cactus native to Arizona and Texas, and every year its flowers bloom all at once—but only for one night. You can see it happen at Tohono Chul botanic gardens in Tucson, Arizona, sometime between the end of May and late July. Sign up for the garden's mailing list to be notified the day of, then hurry down to see the magical Night Bloom event (along with about fifteen hundred other people). The flowers wilt a few hours before sunrise . . . but if you miss them, there's always next year.

Tromsø Arctic-Alpine Botanic Garden—Tromsø, Norway

This is the most northerly botanical garden in the world, which is part of the Arctic University of Norway in Tromsø. Although it's high up in northern Norway, because it's on the coast the area is warmed by the Gulf Stream, so in the summer the weather is temperate and the landscape is lush and green. The garden is home to some otherworldly Arctic, Antarctic, and Alpine plant species. Look for the fragile Arctic buttercups, Greek cushion flowers, or Himalayan blue poppies. You can visit the garden year-round, but come between May and early October if you want to see the flowers in bloom.

FALL IN LOVE WITH (OR IN) THIS OVERGROWN, LEAFY TUNNEL

Tunnel of Love—Klevan, Ukraine

♫♪ Ride down baby, into this tunnel of love ♫♪ You might have already seen this green curiosity on Instagram: It's a train track in Ukraine, overgrown with foliage that passing freight trains conveniently keep manicured in a perfect tunnel shape. Called the Tunnel of Love, it's a twenty- to thirty-minute walk from the small town of Klevan, which is a bus ride from Lutsk in western Ukraine. Once you find the track, take a romantic stroll under a canopy of lush green leafiness for several miles. Trains do still pass through here, though, so be careful.

Travel Tip: Definitely wear insect repellent before you visit the Tunnel of Love, or you'll unfortunately get a whole lot of love from giant mosquitoes.

FOREST BATHE IN A JAPANESE BAMBOO FOREST

Arashiyama Bamboo Grove—Kyoto, Japan

Shinrin-yoku means something like "bathing in the atmosphere of the forest," and it's a term that emerged in Japan a few decades ago to capture something that people in Japan and elsewhere have been practicing for centuries: spending time among trees and nature in order to wind down and recharge. We obviously all need this more than ever, and this bamboo forest in Kyoto is the perfect place to do it, kind of. Here you can walk through a majestic corridor of giant bamboo trees and reflect on your own existence to a soundtrack of birds, rustling leaves, and, if we're honest . . . a lot of chatter from all the other people who came to do the same thing. It can get extremely busy here, but it's still a beautiful place to visit.

Did You Know? There's a hidden bamboo forest in Georgia, USA, too. Just outside of Atlanta, the East Palisades Trail winds along the Chattahoochee River, eventually passing through a peaceful (and less crowded!) grove of towering bamboo trees.

LISTEN TO CLASSICAL MUSIC IN THIS CARIBBEAN RAIN-FOREST GARDEN

Hunte's Gardens—Saint Joseph, Barbados

If you happen to be in Barbados and feel like a break from the beach, drive inland to Hunte's Gardens, a hidden sanctuary in the center of the island. Before the colonial sugar plantations took over the island in the 1600s, Barbados was covered in rain forest, and the garden is teeming with tropical plants and colorful flowers. The owner, Anthony Hunte, plays classical music into the gulley all day long (for the plants!), so it's a serene place to spend a couple of hours—also with a couple of cups of homemade rum punch, if you're so inclined.

Travel Tip: The best way to get to Hunte's Gardens is by car or taxi, but there's also a (sometimes unreliable) bus that leaves from the main terminal in Bridgetown.

DRIVE AMONG UPSIDE-DOWN TREES IN MADAGASCAR

Avenue of the Baobabs—Menabe, Madagascar

Even if you're not a die-hard tree enthusiast, it's hard not to be impressed by the trees of the Avenue of the Baobabs. Alongside a section of dirt road in Madagascar, you'll find a collection of trees that are so surreal looking that it's like walking through an animation. The baobabs (nicknamed "upside-down trees") are more than eight hundred years old, with thick trunks and branches that spread out in a flattened shape at the top, so they look like they were planted with their roots to the sun. You can see the beauties on the dirt road that runs between Andriamena and Marofototra in the Menabe region.

FROLIC AROUND THESE FLOWER FIELDS

Yes, flower fields make for an incredible backdrop, but please: if you're taking photos, don't step on the blooms! Not. Even. One. The destruction you'll cause might seem harmless but can be devastating for the ecosystem and will ruin things for everyone else. Now, with that out of the way: here are four flower fields around the world to admire from the paths.

1. **HITACHI SEASIDE PARK—IBARAKI PREFECTURE, JAPAN**

   The slopes of this stunning park are washed in different colors depending on the season. Visit in October for hot-pink and fluffy kochia bushes or in late April for a sea of baby-blue nemophila flowers that blend in with the edge of the sky.

2. **KAAS PLATEAU—MAHARASHTRA, INDIA**

   Also known as Kaas Pathar, this plateau transforms into a "valley of flowers" during monsoon season. A carpet of delicate, colorful wildflowers covers the landscape, and the color transforms every couple of weeks depending on flowering cycles.

3. **HALLERBOS FOREST—HALLE, BELGIUM**

   This forest is beautiful year-round, but flower lovers should visit around April, when Hallerbos transforms into a purple forest wonderland with a carpet of bluebells. They bloom at different times each year, but you can keep track of their flowering status at hallerbos.be.

4. **WALKER CANYON POPPY FIELDS—LAKE ELSINORE, CALIFORNIA, UNITED STATES**

   Southern California is known for its mysterious "super blooms," but this phenomenon needs a decent amount of rainfall during winter to occur. In 2019 the hillsides of Lake Elsinore's Walker Canyon exploded in a glorious covering of bright-orange poppies; in 2020 and 2021, hopeful visitors weren't as lucky. We know you're reading this in the future, so it's worth checking the city's website (lake-elsinore.org)—along with those of other wildflower hotspots in Southern California—in case another super bloom is coming to a springtime near you.

TAKE A FERRY TO THIS BOTANICAL ISLAND PARADISE

Oedo Botania—Geoje, South Korea

We love a good makeover story, and this botanical garden has completely transformed what was once just a small, barren island of rock off the southern coast of South Korea. The island's paths wind through gardens of fancy topiary, tulips and tropical plants, a cactus section, and a sculpture park. You can take a ferry here from the city of Geoje, and tickets will get you about ninety minutes of wandering the garden's slopes. It's hilly, but there are plenty of spots to sit and take in panoramic views of the green sea and coastal cliffs.

Microdose on Nature: Even if you're visiting a major city, viewing their botanical gardens—most cities have one—or sitting in a leafy park and gazing at the treetops for twenty minutes or so can be enough to help you feel revitalized.

MARVEL AT THE PLANT LIFE ON THIS "ALIEN ISLAND"

Socotra Archipelago, Yemen

Known as Alien Island or the Galápagos of the Indian Ocean, the Socotra Archipelago (about 210 miles south of Yemen) is home to more than three hundred endemic flora species—meaning you won't see these plants growing in the wild anywhere else on Earth. Some plants have names straight out of a storybook (Dragon's Blood Tree; Persian Carpet Flower), and many are so unique and unusual looking (Google "Desert Rose") that you really will feel like you're on another planet. Don't even get us started on all their really cool animals.

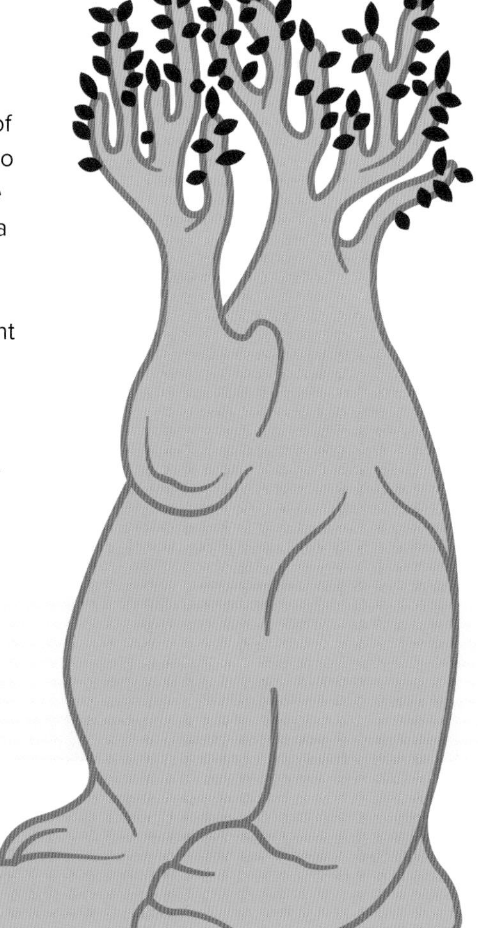

Psst! If you want to read about amazing animal experiences around the world, flip to page 72.

NATURAL HOT SPRINGS

There's something quite magical about hot springs; they're a reminder of just how cool our planet is. Explore these island, desert, and countryside delights—and don't forget to thank Mother Nature for drawing you a wonderful bath.

TRAVEL TO TURKEY'S TRAVERTINE TERRACES

Pamukkale Thermal Pools—Turkey

This village in southwest Turkey has been a hot-springs destination since Roman times—and for good reason. Pamukkale, which literally means "cotton castle" in English, is home to seventeen blindingly white pools that cascade down the side of a ridge. The pools were formed after a spring (with a high volume of dissolved calcium bicarbonate) spilled over the cliff, cooled, and then firmed up, resulting in calcium deposits that became the natural pools you'll see today. After you're done bathing in the magical blue waters, be sure to visit the site of the ancient Roman city of Hierapolis.

Fun Fact: Hot springs are caused by geothermal systems, which cause underground water to warm up and rise to the Earth's surface. They're often found near active (or recently active) volcanoes, where the ground is heated by magma.

TAKE A HOT-SPRINGS TOUR THROUGH A MAGICAL ISLAND

São Miguel—The Azores, Portugal

Located in the middle of the Atlantic Ocean, the archipelago of the Azores was formed thanks to some intense volcanic activity millions of years ago. São Miguel, the biggest island, is the perfect place to go hot springs–hopping. The town of Furnas is bursting with springs; don't miss Poça da Dona Beija Caldeira Velha, a converted taro plantation with a number of pools of varying temps (the hottest one reaches 102 degrees Fahrenheit, or 39 degrees Celsius), or the huge pool at Terra Nostra Park. But maybe the most magical spot of all: Ponta da Ferraria, a hot spring that feeds directly into the ocean. Find a warm spot in the rocks by the hydrothermal vents and swim out into the ocean when you want to cool down. You have to walk down the side of a cliff to get there, but it's absolutely worth it.

SOAK IN A BATHTUB IN THE MIDDLE OF THE DESERT

Mystic Hot Springs—Monroe, Utah, United States

In the middle of central Utah, there's a quirky, rustic little spot where you can soak in natural hot springs—or a vintage bathtub—in the middle of the wilderness. Mystic Hot Springs has two concrete pools along with six tubs, and the temperature of the mineral-rich water ranges from 99 to 110 degrees Fahrenheit (37 to 43 degrees Celsius). It's open twenty-four hours (!), so you can soak as you watch the sunset over the valley or see the stars twinkle above. And if you jump in a tub and end up feeling too relaxed to move, don't worry, there are plenty of on-site accommodation options: think camping and RV sites, converted hippie buses, and cabins.

STROLL THROUGH AN ANCIENT *ONSEN* TOWN IN A *YUKATA*

Kinosaki Onsen—Japan

Not gonna lie, picking just one hot-springs destination in Japan is pretty tough, but this small mountain village about three hours northwest of Kyoto is hard to beat. Founded in 720 CE, this *onsen* town has seven different bathhouses, fed by natural hot springs, that are all within walking distance to one another. It's basically like one big *ryokan* (traditional Japanese inn), and the best part is: it's perfectly acceptable to just wear a *yukata*—a Japanese dressing gown—as you stroll through the streets.

Good to Know: Before you go, it's important to know that *onsens* are a big part of Japanese culture and come with rules and traditions: most baths are separated by gender (although there's a movement pushing for more inclusive, nonbinary practices) and visible tattoos are not generally allowed. Be sure to research the customs or ask a local.

HIT UP AN URBAN NATIONAL PARK WITH FORTY-SEVEN HOT SPRINGS

Hot Springs National Park—Hot Springs, Arkansas, United States

This may be the smallest national park in the United States, but what it lacks in acreage it makes up for in hot springs—forty-seven, to be precise. Nestled in the Zig-Zag section of the Ouachita Mountains, a range that was formed about three hundred million years ago when two tectonic plates collided, this urban park is unlike any other. Here, it's less about hiking (though there are a few short trails) and more about history. And while you can't swim directly in the outdoor geothermal pools themselves, there is an on-site bathhouse that's open to the public. The park also has thermal-spring drinking fountains, so bring along your reusable bottle and sample the water for yourself.

SEVEN FACTS ABOUT VOLCANOES THAT'LL MAKE YOU SAY, "DAMN, THAT'S COOL"

Did you know that more than 80 percent of Earth's surface was created by volcanoes? Who knows where we'd be without these explosive mountains—certainly not soaking in natural hot springs, that's for sure! Let's take a minute to appreciate how incredible volcanoes are with a few fascinating facts.

1. In the most basic sense, volcanoes are openings in Earth's crust that let out gases and molten rock. Think of them like the pores of our planet.

2. Around 1,350 potentially active volcanoes currently exist across the world, excluding the ones on the ocean floor. Of these, approximately five hundred have had an eruption in historical time.

3. Speaking of the ocean, scientists suspect that about 80 percent of Earth's volcanic eruptions occur under the sea.

4. Roughly two-thirds of the world's active volcanoes can be found around the perimeter of the Pacific Ocean in a zone with intense tectonic activity, nicknamed the "Ring of Fire." (It's actually more like a horseshoe shape, but that's nowhere near as catchy.)

5. The biggest active volcano in the world resides on the Island of Hawaii and is more than 13,100 feet (4 kilometers) tall. It's also one of the most active volcanoes, with thirty-three recorded eruptions since 1843.

6. Volcanoes get their name from the Italian island of Vulcano, which was named after the Roman god of fire, Vulcan.

7. While they can be deadly, they also create life: volcanic materials contain high levels of minerals that eventually break down to produce extremely fertile soil.

BATHE LIKE A ROMAN IN THE ITALIAN COUNTRYSIDE

Cascate del Mulino—Saturnia, Italy

Hidden in the Tuscan countryside, about two and a half hours from Rome, you'll find a number of dreamy hot-spring tubs fed by a thermal waterfall, Cascate del Mulino. You can soak in the toasty water (it stays a delightful 99 degrees Fahrenheit, or 27 degrees Celsius, year-round) while you soak up the view—as long as you don't mind the sulfuric smell. As with many beautiful places, it does get busy, so be sure to visit early to avoid the crowds (and *always* properly dispose of your rubbish and treat the location with respect!). The natural pools are free to all, but if you're feeling fancy, there are nearby luxury hotels with private pools, too.

RELAX UNDER A WARM WATERFALL

Spa Thermal Park—Taupō, New Zealand

We'd be remiss not to mention at least one New Zealand spot in this section: the country's location, where two tectonic plates connect, means it has plenty of geothermal activity (and therefore some epic hot-spring action). And Spa Thermal Park, near Lake Taupō in the center of the North Island, has got to be one of the best. Minerally, geothermal waters from Otumuheke Stream meet the Waikato River (the country's longest), creating natural pools and waterfalls that you can relax in. The area is a historical Māori meeting and bathing place, and it's close to the hiking trail that leads to the ever-popular Huka Falls.

Is It Safe to Visit Active Volcanoes? There are risks involved with visiting active volcanoes, even if they're not currently erupting. Before you make any plans, do some thorough research on the location so you're aware of the potential dangers. If you do decide to go, check local advisories for any warnings or closures and consider booking a tour with a reputable company.

GO OFF-ROADING TO A HIDDEN HOT SPRING

Puning Hot Spring—Angeles, Philippines

As they're accessible only in an off-road vehicle, you'll have to jump in a Jeep and drive through a narrow gorge to reach these hot springs. This spa is home to twelve pools, fed by the Sacobia River, which are a delightful 104 degrees Fahrenheit (about 40 degrees Celsius). And after you're done soaking, treat yourself to a heated sand massage, which involves being covered in warm sand and having someone walk (literally walk!) all over your body.

TAKE A DIP IN ICELAND'S OLDEST SWIMMING POOL

Secret Lagoon—Flúðir, Iceland

Everybody knows about Iceland's Blue Lagoon, the famous hot spring known for its milky blue water, but many don't realize it's actually man-made. For an experience that's a little more authentic (and a little less crowded), check out Secret Lagoon, or Gamla Laugin as it's called by locals. This pool has been around for more than 130 years and is apparently the country's oldest. It's fed by nearby bubbling hot springs and a small geyser—Litli Geysir—which you can witness spouting up boiling water on the reg.

WE PROMISE TO STOP TALKING ABOUT VOLCANOES AFTER YOU ADMIRE THESE GORGEOUS CRATER LAKES

Last thing about volcanoes! When a volcano erupts and then collapses, a big depression called a "caldera" can occur. These large holes sometimes fill with water, forming stunning crater lakes that you might like to know about.

1. CRATER LAKE NATIONAL PARK—OREGON, UNITED STATES

Formed 7,700 years ago after the violent eruption of Mount Mazama, Crater Lake is the deepest lake in the United States at 1,943 feet (592 meters). The water is exceptionally blue, and it's also believed by scientists to be one of the purest lakes, thanks to the fact it's fed solely by snow and rain.

2. LAGUNA DE QUILOTOA—ZUMBAHUA, ECUADOR

Hidden in the Ecuadorian Andes, this 1.8-mile- (3-kilometer-) wide caldera is well-known for its brilliant green water. Geologists estimate it reaches depths of around 787 feet (240 meters).

3. SETE CIDADES—THE AZORES, PORTUGAL

Twin crater lakes, one green and one blue, make up the breathtaking Sete Cidades on the Azorean island of São Miguel. For the best vantage point, head to Vista do Rei (King's View).

4. KERID CRATER LAKE—KLAUSTURHÓLAR, ICELAND

At approximately 3,000 years old, Kerid is just a youngin' compared to most other crater lakes in Iceland. And you can tell from its vibrant colors—its slopes are crimson and the water a deep aquamarine, thanks to the relatively fresh minerals deposits. (Actually, you know what? We still have more than a hundred pages to go, so we can't guarantee there'll be no more volcano talk. Sorry.)

ECO EXPERIENCES

It should come as no surprise that tourism can be the absolute worst for the environment. But it can also wreak havoc on the cultures and communities whose lives are often negatively impacted by even the most well-meaning travelers. The good news is that every year, more and more amazing eco-conscious experiences—from the basic to the downright luxurious—are popping up as the demand for more ethical and environmentally friendly travel increases.

WHAT IS ECOTOURISM, ANYWAY?

The International Ecotourism Society defines it as "responsible travel to natural areas that conserves the environment, sustains the well-being of the local people, and involves interpretation and education." And while it might seem that these days everything is "eco" something, when searching for ecotourism destinations and experiences, look for a combination of the following:

- activities that have a low environmental (and social) impact
- buildings that use sustainable and natural materials and fibers and green energy sources
- ownership and management by, or partnership with, Indigenous or local communities
- a commitment to reinvesting the profits back into the community or environmental conservation or both
- dedication to raising cultural and environmental awareness

You can read more about the principles of ecotourism at ecotourism.org.

EXPLORE SACRED LANDS IN THE ECUADORIAN AMAZON

Napo Wildlife Center—Yasuní National Park, Ecuador

Napo Wildlife Center is a luxurious eco-lodge in Yasuní National Park, a UNESCO-listed Biosphere Reserve in the Ecuadorian Amazon. Situated on the banks of Anangucocha Lake, the lodge is accessible only by boat, and once you're there you can explore rain-forest trails, spot toucans and more incredible wildlife, learn about ancestral customs, and relax in the beautiful jungle suites. The reserve is managed by the Indigenous Kichwa Añangu community, who reinvest the proceeds back into projects that support the community and environmental preservation.

STAY AT AN ORGANIC FARM AND ECO-LODGE IN COSTA RICA

Villas Mastatal—Mastatal, Costa Rica

Twenty-five percent of Costa Rica is protected conservation area, so this small country has a lot on offer in the way of ecotourism. Here's one for your radar: Villas Mastatal, a family-owned eco-lodge near the small village of Mastatal in the country's Central Valley. You can sleep in villas with jungle views, perfect your Chaturanga on the yoga platform surrounded by nature, try to spot a sloth on a hike in the nearby Cangreja National Park, or pass the afternoon like a sloth yourself in a hammock. The property also has a permaculture farm that you can help out on so you can get your hands a little dirty in between some serious relaxation.

FIVE ECO-LODGES IN THE UNITED STATES AND CANADA

1. **SHASH DINÉ ECO-RETREAT—PAGE, ARIZONA**
   Experience natural beauty on Navajo land in northern Arizona, with a glamping-style B&B and a ranch that offers a unique Navajo experience.

2. **WILDSPRING GUEST HABITAT—PORT ORFORD, OREGON**
   A small eco-friendly resort where you can fall asleep to forest sounds and then wake up to have breakfast by the ocean.

3. **CREE VILLAGE ECO LODGE—MOOSE FACTORY ISLAND, ONTARIO**
   Located in the Moose Factory Island community in the James Bay region. Visit in summer or winter for plenty of outdoor adventures.

4. **SADIE COVE WILDERNESS LODGE—KACHEMAK BAY STATE PARK, ALASKA**
   A family-run beachfront lodge in Kachemak Bay that operates entirely on alternative energy sources and is a great starting point for those looking to explore the wilderness.

5. **SPIRIT BEAR LODGE—SWINDLE ISLAND, BRITISH COLUMBIA**
   A cozy lodge in the heart of the Great Bear Rainforest owned by the Kitasoo/Xai'xais people, the First Nations community of Klemtu on Swindle Island.

TREK WITH GORILLAS IN RWANDA

Volcanoes National Park—Musanze, Rwanda

Gorilla trekking brings significant income to local communities in Rwanda, and visiting Volcanoes National Park in the country's northwest means you're contributing to the conservation of these endangered beauties. The park is home to ten gorilla families and is committed to their well-being: only eight people are allowed to visit a gorilla group per day, and each tour group can stay with them for only one hour. The permit is pricey, but it's worth it, and aside from gorillas you can spot adorable golden monkeys and other wildlife in the park.

Travel Tip: If you're looking for accommodation in the area, check out the luxurious Forest Villas at Bisate Lodge, just a short drive from the park's headquarters.

GLAMP IN A SWEDISH WINTER WONDERLAND

Sápmi Nature Camp—Laponia, Sweden

Sápmi Nature Camp is a glamping experience in the UNESCO World Heritage–listed Laponia, founded by a member of the Indigenous Sámi community in the Sjavnja nature reserve. Guests stay at the camp to learn about Sámi culture, help herd reindeer, and explore the magical snowy landscapes using traditional wooden skis. You'll sleep in cozy tepee-style *lavvu* tents and help prepare traditional Sámi cuisine with locally sourced fish, meat, herbs, and berries. It also feels pretty personal—only ten guests can stay at the camp at one time.

You Now Need an Eco-Visa to Visit This Country: The tiny island nation of Palau in Micronesia is so committed to preserving its culture and natural environment that in a world first, the government has actually updated their immigration policy to include a mandatory eco-visa for all visitors. The children of Palau wrote a pledge that every tourist has to take before passing through customs, committing to help keep the island beautiful for generations to come. There's an extremely cute and emotional video explaining it all; you can watch it and learn more here: palaupledge.com.

SUPPORT RAIN-FOREST PRESERVATION IN GUYANA

Iwokrama Forest—Guiana Shield, Guyana

In 2020 Guyana was named as the world's best ecotourism destination at the ITB travel and trade show in Berlin, and there's no end to the beauty you can explore in this corner of South America. Eighty-seven percent of the country is covered in rain forest, like the Iwokrama Forest, one of the four great pristine rain forests in the world. It's a protected area of almost one million acres with more biodiversity than you can poke a hiking stick at (but don't do that!). It's also home to the Indigenous Makushí people, and profits from tourism here are reinvested back into the community.

VISIT A DESERT MUSEUM IN INDIA

Arna Jharna—Moklawas, Rajasthan, India

Arna Jharna is a "living museum" and desert treasure that celebrates the surrounding landscapes of the Marwar region of Rajasthan in northern India. Visitors can expect an interactive experience that includes learning about the area's biodiversity, rural practices, and the site's own water-harvesting system. Inside the red-earth museum you'll find an extensive collection of traditional and ritualistic brooms, musical instruments, and pottery. The surrounding area is rich with desert flora and fauna, and you can see peacocks and deer wandering the area. It's a perfect place to visit on a day trip from Jodhpur.

FOUR TIPS FOR ECO-CONSCIOUS PACKING

It's true that the bulk of the damage to the environment is coming from big corporations (and, unfortunately, all those flights contribute as well), but it's also true that all of our smaller choices add up. Here are a few more small ways to reduce your own environmental impact, because being an eco-conscious traveler starts before you even leave your home.

1. **Pack an ocean-friendly sunscreen.** Do you know how much sunscreen washes into the sea each year? It's a lot, and it causes coral bleaching and other damage to marine life.

2. **Bring your own meal kit.** Pack your own spork and a cloth napkin so you don't have to use disposable utensils and napkins for those delicious street eats. Don't forget the reusable water bottle!

3. **Ditch the travel-size toiletries.** Those tiny versions of your favorite products are convenient as hell, but think of all the plastic. Buy a set of refillable containers (or reuse small jars and tins you already have!) and load them up with your toiletries before you leave.

4. **Stash a reusable tote bag.** That's one less plastic bag you'll need to ask for when you're souvenir shopping and one small step toward a greener vacation.

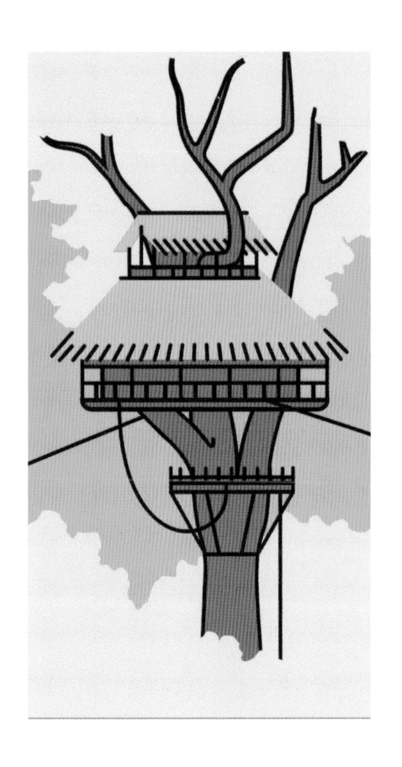

ZIP-LINE OVER LUSH FOREST IN LAOS

The Gibbon Experience—Nam Kan National Park, Laos

The Gibbon Experience is a forest-based conservation project in Bokeo Province in northwest Laos. They're known for their network of zip lines and tree houses (the highest in the world!) that you can book to stay in if you want to wake up to what might be the best treetop views in the world. Guided hikes will take you into the forest for gibbon spotting, but even if you don't see the apes, the scenic treks through the national park and zipping from mountain to mountain are worth visiting for alone. And by visiting, you'll be helping to sustain the livelihoods of the communities that live in the forest.

SEE TURTLES ON A NIGHT WALK IN MEXICO

Centro Ecologico Akumal—Yucatán Peninsula, Mexico

Akumal is Mayan for "place of the turtle," and snorkeling with green turtles in this part of Yucatán is a highlight for many. But part of what makes this possible is the work of Centro Ecologico Akumal (CEA), an organization committed to turtle conservation and preserving the beautiful coastal and marine ecosystems in the area. Aside from taking an opportunity to volunteer, you can make a donation to the CEA and sometimes join night walks to see mother turtles or their babies hatching or both. Like anything this special, it's not guaranteed, but if you're staying in Akumal, it's definitely worth inquiring about.

BIZARRE BEACHES AND BAYS

There are few things that say "vacation" more than lounging on a beautiful beach—and don't get us wrong, that sounds great—but you can do that anywhere. If you're looking for something more memorable, here are a few beaches, bays, and shores that have something different to offer, from colorful sand to glowing water.

SEE A FORMER TRASH SITE TURNED GLISTENING GLASS BEACH

Glass Beach—Fort Bragg, California, United States

For more than half a century, until 1967, this coastal California spot in Fort Bragg was the city's trash dump, littered with everything from bottles to broken appliances and used cars. The practice of disposing of garbage here stopped (thankfully!!), and nature took matters into its own hands, slowly eroding the waste into small pieces. Though it was once trash, this colorful "sea glass" is now what attracts people to the area.

Travel Tip: Visit Glass Beach at low tide for the best sea-glass visibility.

SWIM WITH WILD PIGS

Pig Beach—Big Major Cay, Bahamas

Many people visit the Exumas, a chain of islands in the Bahamas, for the crystal-clear water and whiter-than-white sand beaches. But there's another reason why tourists (and celebs) make the trip to Big Major Cay, a tiny island uninhabited by humans: for the chance to swim with pigs. That's right, a group of feral hogs are the only locals you'll find here; how exactly they ended up at the beach is a bit of a mystery, though. You can visit Pig Beach on a tour or by renting your own boat. If you do visit, remember these are wild pigs and this is their home, so please be respectful. And watch out—these pigs can bite.

STROLL ALONG A RARE SHELL-PACKED SHORE

Shell Beach—Shark Bay, Australia

If you're one of those people who love going to the beach but hate finding sand in their bag/car/ear for days afterward, this one's for you. Trillions (yes, with a *T*) of little white shells replace sand at this Shark Bay beach that extends for seventy-five miles. They all come from one animal—the Hamelin cockle—that inhabits the hyper-salty water along with very few natural predators. The shells on the beach are about thirty-two feet deep, and they contrast beautifully with the brilliant blue water.

FIND MEXICO'S HIDDEN BEACH INSIDE A CAVE

Playa del Amor—Marieta Islands National Park, Mexico

Playa del Amor, or Hidden Beach as it's otherwise known, is easy to spot from above; a single hole in the island's otherwise lush vegetation makes it look like a putting green. From down below, you'll feel as though you're swimming in an underground crater, looking up through a hole in the Earth's crust. This magical beach is found on Isla Redonda (Round Island), one of two uninhabited islands that make up this archipelago off the coast of Puerto Vallarta. Unfortunately, though, it's not exactly a secret anymore. In order to protect the environment from overtourism, visitors are limited to around one hundred per day, so be sure to book ahead.

GO DRAGON SPOTTING ON A PINK-SAND BEACH

Pantai Merah—Komodo Island, Indonesia

The most obvious reason to visit Komodo Island is, well, to see Komodo dragons—a.k.a. the world's largest lizards—in their native habitat. But these enormous reptiles aren't the only drawcard: the island is also home to a rare pink beach called Pantai Merah. The beach gets its hue from small fragments of red coral that mix with the sand, creating a pinkish shoreline that contrasts magically with the turquoise waters. The island forms part of Komodo National Park, so there is an entry fee, and if you go be sure to practice responsible tourism to protect wildlife and the environment. Oh and seriously, be careful of the dragons: they're carnivorous and have a poisonous bite!

FLOAT THROUGH A GLOW-IN-THE-DARK BAY

Mosquito Bay—Vieques, Puerto Rico

Our planet has a handful of beaches and bays that experience bioluminescence, but none shines as bright as Mosquito Bay in Puerto Rico—and it has the Guinness World Record to prove it. Microscopic organisms glow a neon blue when disturbed, making the water look like it's swimming with shining stars. The phenomenon occurs here year-round, though it looks better the darker it is, so avoid visiting during a full moon if you can. To fully experience the magic, hop in a clear-bottom kayak and watch the water light up as you paddle. For conservation reasons, no swimming or motorboats are allowed in the bay, and it's best to skip the bug spray as it may harm the plankton.

BASK ON A BLACK SAND BEACH

Playa Jardín—Tenerife, Spain

On the largest of the Canary Islands, in the resort town of Puerto de la Cruz, you'll find a beach that glistens with black volcanic sand. Named Playa Jardín (or Garden Beach), it'll come as no surprise that the beach backs on to a lush garden where plants and trees add colorful pops of green to the landscape. This beautiful setting was no accident, though—it's one of three beaches designed by local artist César Manrique. Breakwaters have been installed to make swimming a little more relaxing, but the waves can still get pretty wild the farther out you go, so proceed with caution.

HIKE TO ONE OF THE ONLY GREEN-SAND BEACHES IN THE WORLD

Papakolea Beach—The Big Island, Hawaii, United States

Yep, more colored sand! Papakolea Beach, or Green Sand Beach, is a secluded shore on the southeast coast of Hawaii. It sits at the base of Pu`u Mahana, a volcanic cone formed by magma rich in olivine, a semiprecious stone formed in high-heat environments. This is what gives the beach its hue (FYI: it's not *green* green, but it's still cool to see). You'll need to hike down about two miles to reach it, but we promise you won't regret it.

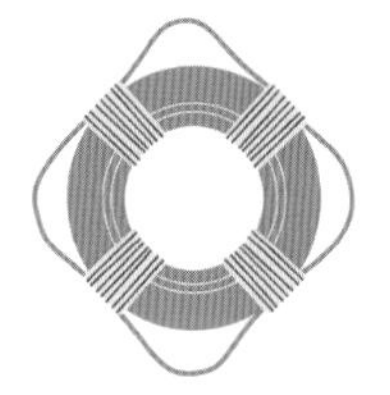

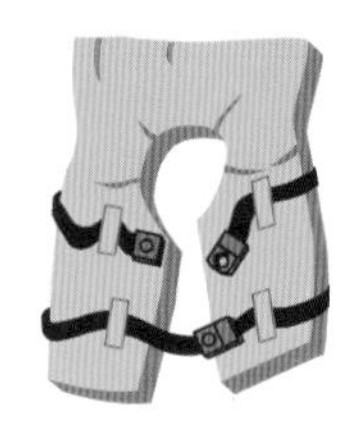

FIVE BEACH SAFETY TIPS EVERYONE SHOULD KNOW

The ocean is a beautiful, powerful force—but it doesn't mess around. Here are a few things to keep in mind before you dive in.

1. It has to be said: Protect yourself from harmful rays by donning a hat, sunblock, and sunnies.

2. Read (and obey) the signs and listen to lifeguards. If a beach says "No Swimming," there's probably a good reason for it!

3. Be careful of ocean currents, especially riptides that can pull you out to sea. If you get caught in a rip, try to stay calm, don't attempt to fight it, and signal for help.

4. Don't swim on your own. Always swim with a mate or make sure someone is looking out for you.

5. Know your limits. If you're not a strong swimmer, don't venture out into uncharted waters.

EXPLORE A SUN-LIT SEA CAVE

Benagil Cave—Algarve, Portugal

A beach inside a massive sea cave with a natural built-in skylight? You have our attention. Located in Algarve on a coastline with countless amazing swimming spots, Benagil Cave has to be up there as one of the most incredible sights in all of Portugal. The fierce waves of the Atlantic carved this limestone cave from the base, while rainfall caused the roof to erode, leaving a big hole that lets sunlight drench the hidden beach inside. The only way you can access the inside is by boat, so do yourself a favor and book a local tour or brave the waters on a stand-up paddleboard or sea kayak.

DISCOVER THE RED SAND BEACHES OF IRAN'S RAINBOW ISLAND

Hormuz Island—Hormuz, Iran

A boat ride away from mainland Iran in the Persian Gulf, Hormuz Island is best known for its colorful soil; it's home to more than seventy different colored minerals. It's a popular spot for artists, who are drawn to the vibrant, multihued landscape, so expect to see murals painted along the streets. But perhaps one of the coolest spots is Red Beach, a coastline made up of crimson, iron oxide–rich soil that contrasts spectacularly against the blue sea. While you're there, spend some time exploring (and taking way too many photos of) Rainbow Valley.

AMAZING ANIMALS

The Earth is home to an estimated 8.7 million different animal species, and, amazingly, the majority haven't even been documented yet. An alarming number of animal species now face extinction thanks to human activity, so if your travel bucket list includes encounters with wildlife, it's important (but also easy!) to make sure whatever you're doing won't harm the animals, alter their behavior, or destroy their habitat. Look for conservation and education over entertainment and exploitation. Now let's dive in! Here are some of the incredible wildlife encounters you can have around the world.

FLOAT WITH MILLIONS OF GOLDEN JELLYFISH

Jellyfish Lake—Southern Lagoon, Republic of Palau

You can float like a jellyfish—*with* jellyfish—in this surreal snorkeling experience. Jellyfish Lake is an enclosed lake in Palau, a tiny island in the western Pacific Ocean, and over millennia the millions of jellies who call the lake home have evolved to lose their sting. Now you, reader, can safely snorkel through a turquoise lagoon among the golden jellyfish as they float through the water following the sunlight. The lake took some time out to recover its gelatinous population a few years ago, but at the time of writing it's open and ready for visitors.

Psst! For more on the extremely eco-friendly Palau, flip to page 61.

BIRDWATCH AT THIS FLAMINGO PARADISE

Lake Natron—Northern Circuit, Tanzania

This lake is an important hatching ground for millions of East African flamingos—but it's also a mysterious force that turns other creatures to stone. So what's going on? The lake has such a high salt and soda content that it has become a deadly brine, and while flamingos have adapted and now flourish in this salty goodness, other birds that accidentally take a dip are calcified as they dry. The lake is also scalding hot, so to recap: don't go in the lake unless you're a flamingo. But definitely walk around it if you want to see some.

GO KAYAKING WITH BELUGA WHALES

Churchill River—Manitoba, Canada

They call it the canary of the sea, and when you get up close to a beluga whale and hear its adorable high-pitched chirps, you'll know why. After the ice breaks and the water warms, these incredible creatures make their way to Churchill River in Manitoba, where you can view them from the shore or, if you want a closer look at their adorable faces, by Zodiac boat or kayak. There are a few different tours available, but Sea North Tours is worth looking into for its commitment to beluga-friendly boat practices. Visit seanorthtours.com for more info.

PARTY AT THE ANNUAL SHEEP ROUNDUP IN ICELAND

Réttir—All over Iceland

You might count sheep to fall asleep, but head to the annual sheep roundup in Iceland and you can do it in real life. *Réttir* is a cultural tradition that happens every September, when farmers nationwide need to bring the woolly squad home after an indulgent summer grazing in the mountains and valleys. Everyone lends a hand to herd the sheep across the hills and into special rings called *retts* to be sorted so every last one can be returned to the right farmer. And after the sorting, it's time to eat, drink, and *baaa* merry all night long at the Réttaball after-party.

Travel Tip: Around Svartárdalur Valley, Skagafjörður and Eyjafjörður in northern Iceland are great places to see the *réttir*.

MEET VICUÑA, LLAMA'S FLUFFY, GOLDEN COUSIN

Reserva Nacional Pampa Galeras Barbara D'Achille—Lucanas Province, Peru

If you think llamas are cute, wait until you meet their relatives. The Pampas Galeras National Reserve in Peru was established as a sanctuary for vicuñas, small camelid creatures with coats of soft golden fleece. They were once endangered (having been hunted relentlessly by Spanish conquistadores) but are now thriving thanks to the reserve and a pre-Hispanic tradition called *chaccu*. It's a ceremonial herding of the vicuñas in which villagers come together to round up the fluffy beasts and humanely shear their coats of their very valuable wool. The practice of *chaccu* discourages poaching, benefits the local communities economically, and is accompanied by a three-day festival so the vicuñas get to live in abundance and everyone else gets to have a good time. The *chaccu* in Pampas Galeras happens every year on June 24.

SNORKEL WITH NEW FRIENDS AT SHARK RAY ALLEY

Hol Chan Marine Reserve—Near San Pedro, Belize

Did you know that not all sharks want to eat you? At Hol Chan Marine Reserve in Belize, you can swim with nurse sharks, who are generally a chill bunch as long as you don't feed or harass them. Shark Ray Alley is just south of the Hol Chan cut—a shallow diving spot known for clear waters and colorful reefs. Visit this poppin' alley with an ethical tour and you can snorkel with the friendly sharks, sting rays, and even sea turtles. You'll definitely want to bring an underwater camera!

Travel Tip: When you're choosing an ethical tour guide, check the reviews and avoid any that mention touching animals, feeding them, or holding them for photos.

SEE MORE THAN KANGAROOS ON THIS AUSSIE ISLAND

Kangaroo Island, South Australia, Australia

If you're going to Australia to see animals, this island is a must-visit. It's named after the tens of thousands of kangaroos that live there, but there's so much other wildlife on the island that it's known as a "zoo without fences." Think: koalas, wombats, echidnas, and sleepy sea lions on sand dunes. And there's plenty more to do: food and wine tours in rolling green countryside, adventure activities, or swimming in the turquoise waters of the pristine white-sand beaches. Kangaroo Island was devastated by the 2019–2020 Australian bushfires, the largest in its recorded history. Since 2020, the island has been in recovery, and visiting this beautiful place will support the local communities in their efforts to rebuild.

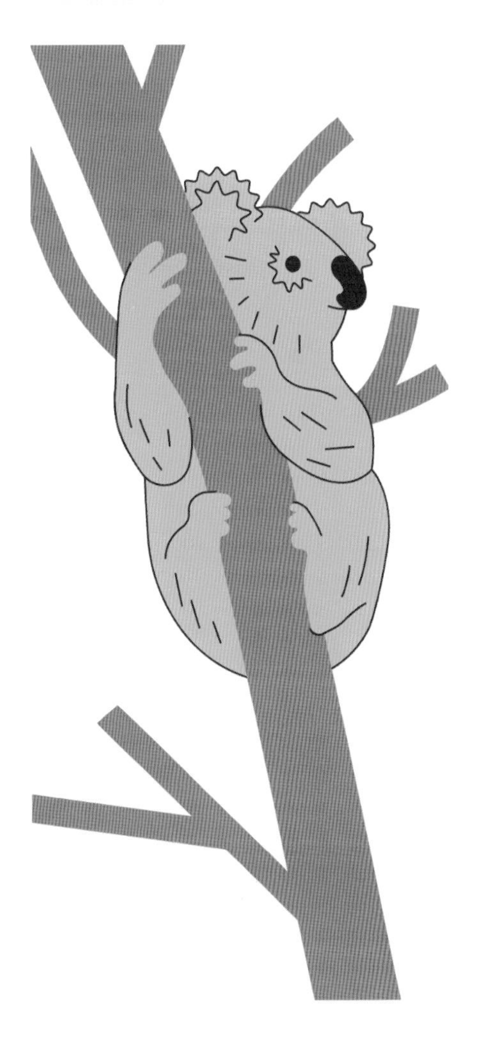

ROTTNEST ISLAND, AUSTRALIA

Getting the Perfect Quokka Selfie

If you haven't heard of quokkas, they're basically like tiny chunky kangaroos, and they always seem to be smiling in photos. If you visit Rottnest Island in Western Australia, you might be lucky enough to get a selfie with one! Here are some tips for getting the perfect pic:

1. The first rule: never feed the quokkas human food. It will make them sick. This applies to all wildlife, which you now already know.

2. Get down on the ground and let a quokka come to you.

3. No sudden movements; use soft voices.

4. Wait for the creature to pick up a leaf (or seed or root) and start munching. Hold out your camera (selfie stick optional!) and try to capture a pic while the quokka has its tiny mouth open.

5. Thank the quokka for its service and move on.

Ready to meet one? See more of Rottnest Island and its adorable quokkas on Instagram: @rottnestislandwa.

WALK ON WATER WITH A FRENZY OF FISH AT YOUR FEET

Isdaan Floating Restaurant—Talavera, Philippines

Isdaan is a Thailand-inspired "floating resto-fun park" with a fish walk that alone makes the visit worth it. After you're done eating (mostly traditional Filipino food) in the restaurant's "floating" bamboo huts, it's time to kick off your shoes and take the fish for a walk. Grab some of the fish food provided, stroll along the long and shallow water walkway, and dozens of hungry koi will soon follow you, nibbling your toes and groveling at your feet for flaky treats. It might make you squeal at first, but you get used to it.

Good to Know: Isdaan has three locations, but visit the one in Talavera if you want to do the fish walk.

SUPPORT CONSERVATION EFFORTS IN KENYA

Lewa Wildlife Conservancy—Isiolo, Kenya

If seeing the Big Five in Africa is on your bucket list, consider visiting the Lewa Wildlife Conservancy at the foothills of Mount Kenya for a conservation-focused safari experience. Together with the neighboring Borana Conservancy, Lewa has created a huge rangeland for endangered black rhinos, herds of elephants, and the world's largest population of Grevy's zebras, along with giraffes, lions and cheetahs, buffalo, and birds. Lewa is focused on sustainable tourism with a minimal environmental footprint and works on dozens of initiatives with local communities to ensure that its sanctuary benefits everyone's livelihoods.

Good to Know: Botswana is also known for its eco-conscious (but very luxurious) safari experiences. Check out Okavango Delta or Chobe Game Lodge.

TAKE A TRIP TO MONKEY ISLAND, PUERTO RICO

Cayo Santiago, Puerto Rico

There's a small, leafy island off the coast of Puerto Rico that is owned and controlled by about two thousand rhesus macaque monkeys, and with the proper documents, the monkeys will allow you to visit. They're studied by the Caribbean Primate Research Center at the University of Puerto Rico (which owns the land), and there are special cages set up on the island for the humans to hide in when the monkeys get too intense. Visits to the island can be arranged only with the research center, and you'll need a special medical clearance to keep the monkeys (and yourself) safe. Visit their website for more info, at cprc.rcm.upr.edu.

GEOLOGICAL WONDERS

Who needs to escape into sci-fi movies or fantasy novels when Earth is this awe-inspiring? Sandstone waves, giant sinkholes, forests made of stone . . . It's hard to believe these natural formations are even real—but they are. Promise.

MARVEL AT THE MARBLE CAVES

Capillas de Mármol—Patagonia, Chile

Patagonia has no shortage of natural wonders, but the Capillas de Mármol, or Marble Caves, have gotta be up there in first place. Sculpted by wind and water over thousands of years, these limestone chambers can be found on Lake General Carrera, a remote lake in southern Chile. Also known as the Marble Chapels, these formations are revered for their curved walls, intricate columns, and swirled ceilings. They're stunning at any time of day, but sunrise and sunset make for an extra-special sight, when the dappled light reflects off the brilliantly clear water and into the caves.

SEE THE OCEAN "DRAIN" INTO A NATURAL HOLE

Thor's Well—Cape Perpetua, Oregon, United States

Just south of Cape Perpetua on Oregon's jagged coastline, there's an opening that appears to suck the ocean right up. This spectacular formation, known as Thor's Well, was likely once a sea cave until its roof collapsed, leaving a gaping hole that looks almost bottomless (though it's estimated to be only around twenty feet deep). Depending on the tide, you might see the water gush into the hole or spurt up to forty feet into the air, and while it can be spectacular during a storm it can also be very dangerous, so make sure to obey the signs and stay a safe distance away.

HIKE ALONG A SANDSTONE WAVE

The Wave—Coyote Buttes North, Arizona, United States

If you've seen photos of the southwestern United States, chances are you've seen the geographical wonder known as the Wave. Located in Coyote Buttes North, an area in the north of Arizona just over the border from Utah, this spectacular rock formation looks like a natural half-pipe, rippled with lines of pink, orange, red, and white Navajo sandstone. It was formed by water and wind erosion over millions of years. You'll have to hike six miles to get there, though, and score a coveted permit, so make sure you plan ahead.

Pro Tip: The best time to snap a pic is midmorning to early afternoon, when it's not too shadowy (but TBH it's quite hard to take a bad photo here).

FLY OVER A MAZE OF SANDSTONE DOMES

Bungle Bungle Range—The Kimberley, Australia

This hidden gem in Purnululu National Park, part of Australia's remote Kimberley region, was forged more than 350 million years ago. It's been an important site for Aboriginal Australians for more than 40,000 years, but remained mostly secret from the rest of the world until the early '80s. Formed by a dissolving bedrock, it's a maze of black-and-orange-striped domes that resemble giant beehives. From the ground it can be hard to take in the scale of these rocks, which stretch up to 820 feet (250 meters) into the sky, so for the best views opt for a scenic flight.

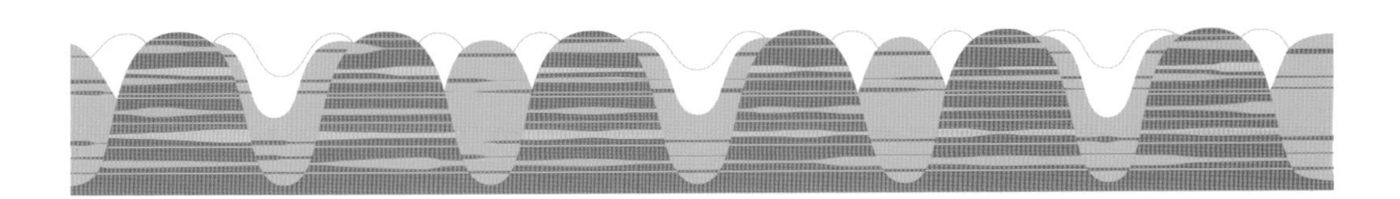

SWIM IN MEXICO'S STUNNING CENOTES

Dos Ojos—Yucatán Peninsula, Mexico

Sure, Mexico's beaches are beautiful, but for a more unique experience, skip the sand and head to one of the Yucatán Peninsula's hidden swimming holes known as cenotes. These deep sinkholes, formed by the collapse of limestone rock, have natural pools filled with crystal-clear water at the bottom. They can be open, partially covered, or like a full-on cave, but they're idyllic in any form. Cenotes are sacred in Mayan culture, and there are thousands sprinkled along the peninsula. But if you only have time to snorkel (or dive) in one, make it Dos Ojos (which means "two eyes"). Located near Tulum, this underwater cave system spans eighty-three kilometers, making it one of the biggest in the world.

SCUBA DIVE FOUR HUNDRED FEET DOWN INTO EARTH'S CRUST

Great Blue Hole—Belize

Just when you thought we were done talking about sinkholes . . . Around 50 miles (80 kilometers) off the coast of mainland Belize lies the aptly named Great Blue Hole, a giant hole in the ocean floor that stretches 1,000 feet (300 meters) wide and shoots down more than 400 feet (120 meters). It's so big that it's even been photographed from space! It's popular with divers, for obvious reasons, and though the surrounding reefs are full of life, there isn't much going on at the bottom of the hole—the lack of light and oxygen makes it rather unlivable. Wondering how we ended up with this beautiful oceanic pit? It remains a bit of a mystery, but scientists think it probably formed during the last Ice Age.

VISIT ONE OF THE MOST INHOSPITABLE PLACES ON EARTH

Danakil Depression—Afar, Ethiopia

Okay, we probably didn't totally sell this one with the title, but hear us out: Up in the north of Ethiopia sits a desert plain that's truly like nowhere else on Earth. The alien-like landscape is home to active volcanoes, colorful hot springs, geysers, vast salt plains, and mineral-rich lakes and pools, and much of the area is dotted with brightly colored deposits (think fluorescent greens, yellows, and oranges) thanks to sulfur and potassium salts. Temperatures in the area have been known to soar above 112 degrees Fahrenheit (50 degrees Celsius), making it one of the hottest—and most inhospitable—places in the world. It's long been an important location for locals, who come to farm the salt, but in recent years tours have been offered for intrepid travelers.

FLOAT IN A BALLOON OVER FAIRY CHIMNEYS

Göreme—Cappadocia, Turkey

Underground cities, fairy chimneys, hot-air balloon rides—the region of Cappadocia is magical for many reasons. This landscape was formed after nearby volcanoes erupted and covered the area with tuff (soft rock), which was carved away over millions of years by rain and wind, leaving the fairy tale–like chimney rock formations (also called hoodoos) you can see today. There are plenty of ways to explore, like hiking in Göreme National Park, sleeping in a cave hotel, or touring the ancient underground cities. But if you ask us, nothing beats taking flight in a hot-air balloon at sunrise so you can appreciate the beauty from above.

HIKE THROUGH A HIDDEN CANYON

Stuðlagil Canyon—Jökuldalur, Iceland

The phrase "jaw-dropping" gets thrown around a bit these days, but we're putting it out there: when you lay eyes on this spectacular basalt column canyon, you may find your mouth literally falls open and leaves you gaping in disbelief. Thousands of hexagonal basalt columns, formed by slowly cooling lava, tower over the bright-turquoise water of the River Jökla. This wonder was a secret until relatively recently: it used to be completely covered by the river, but when a hydroelectric plant was built nearby, it caused the water level to drop significantly and revealed the canyon.

ENTER THE BIGGEST ICE CAVE IN THE WORLD

Eisriesenwelt—Werfen, Austria

This magnificent labyrinth cave began forming millions of years ago and winds for 25 miles (40 kilometers) through a mountain in the Austrian Alps. The first half mile or so is covered in glittering ice formations, like frozen stalagmites and waterfalls, caused by water that seeps through fissures in the limestone rock and freezes in the cooler parts of the cave. To reach the entrance, you'll have to trek 400 feet up the side of the mountain—or brave a ride on the steepest gondola lift in Austria. There's a decent amount of stair climbing required once you enter the cave, too, but at least you won't have to worry about overheating: the temperature inside usually hovers around a frosty 32 degrees Fahrenheit (0 degrees Celsius).

EXPLORE A FOREST MADE OF STONE

Shilin—Yunnan Province, China

In an area in southern China once covered by an ancient sea, spectacular limestone rock formations appear to grow out of the ground. Shilin (a.k.a. the Stone Forest) is one of the most unique examples of karst, a type of landscape caused by limestone erosion, anywhere in the world. Some pillars stretch up to 160 feet high, while other stones are rounded like mushrooms or sharp and pointy like swords. Designated a UNESCO Global Geopark, Shilin is an important spot for scientists and the Indigenous Sani people of Yi, who have called it home for more than two thousand years.

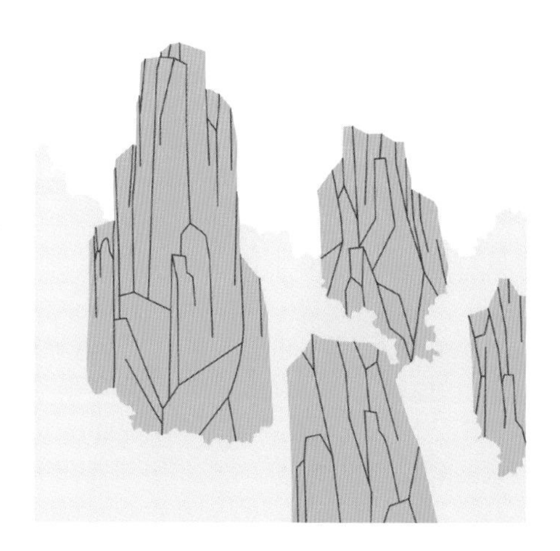

SURREAL SKIES

Star parties and sunsets, moonwalks and rainbows. The sky's a marvelous limit when it comes to the number of incredible wonders up above, so remember to look up when you're traveling.

EXPERIENCE THE MIDNIGHT SUN IN ICELAND

Grundarfjörður, Iceland

For those of us who take it for granted that the sky will always get dark at night, the midnight sun really is an eerie experience. Also known as the polar day, the phenomenon happens during the summer months in areas close to the Arctic and Antarctic Circles. One place to see it happen is Grundarfjörður, Iceland. The picturesque town is right next to Mount Kirkjufell (Iceland's most photographed mountain), and aside from being the perfect place to catch rays at midnight, it's also an amazing area to explore for its natural beauty and wildlife.

So . . . does a "midday moon" exist in Iceland? Maybe, but you'll have to go farther north to experience a true polar night—meaning twenty-four hours of darkness. Iceland still gets a few hours of sunlight a day during the winter.

CAMP BENEATH THE STARS IN TAIWAN

Taroko National Park, Taiwan, China

Nature lovers, if you haven't heard of Taroko Park, put down this book and look it up now. It's a huge, ecologically diverse natural wonderland with giant lush mountains, hiking trails, fir forests, rolling green hills, a famous scenic gorge, and stunning blue-green creeks. It's also home to Hehuan Mountain, the first Dark Sky Park in Taiwan and the best place on the island to gaze up at the Milky Way. Hike the mountain and pitch a tent for a stargazing experience you will never forget.

CHASE ACTUAL RAINBOWS IN HAWAII

Island of Oahu—Hawaii, United States

You've probably seen a basic rainbow, or maybe even a mythical "double rainbow," but did you know that even then you're only seeing part of the picture? Visit the Aloha State, and you can see entire *circles* of rainbow—rainbows in their full form. Rainbows have an important cultural significance in Hawaii, and with its cumulus clouds and clean air, it's also one of the best places in the world to see them. 'Bow enthusiasts can take a helicopter tour over Oahu with Rainbow Helicopters to see all kinds of colorful sky action, including the iconic Full Circle rainbows, if the conditions are right. Check out some photos (if you don't mind spoilers) on their Instagram: @rainbowhelicopters.

Have You Heard of Ghost Rainbows? They're actually called fogbows, and they happen when sunlight mixes with fog to create a rainbow-like arch—only instead of being colorful, it's white. If you ever find yourself in a gentle fog (like in New Zealand or San Francisco) but the sun is still shining, look around and you might just be lucky enough to catch one!

OGLE EPIC THUNDERSTORMS (FROM A SAFE DISTANCE)

Grand Canyon North Rim—Coconino County, Arizona, United States

You might have heard about the South Rim's sunsets, but head to the "other side" of the Grand Canyon for something entirely different. During the summer rainfall season, the North Rim offers an incredible vantage point to watch wild thunderstorms as they approach from a distance. Of course, lightning is a real and serious danger if you're outside, so it's safer to watch a thunderstorm near a building (check out the North Rim's Grand Canyon Lodge). Definitely check the weather forecast for storms before you go hiking. As the website at nps.gov says: "When thunder roars, go indoors."

STAND OVER A SEA OF CLOUDS IN FRANCE

Pic du Midi—Pyrenees, France

Standing beneath the clouds is fine, but standing above a sea of clouds is an exhilarating experience, and one of the best places to do this is at the Pic du Midi Observatory in the French Pyrenees. A cable car will take you for a fifteen-minute ride up to the observatory, which at 9,349 feet (2850 meters) will give you impressive views of the surrounding mountain ranges as the clouds hang below you like a giant marshmallow rug. If you're brave, step out onto the Pontoon in the Sky—a metal walkway that juts out with a glass floor at the end for extra thrills. Stay on after dark (with a reservation) if you want to stargaze; Pic du Midi is the only Dark Sky Reserve in France, and it's home to some pretty powerful telescopes.

WATCH THE SUN RISE AND SET FROM THE SAME PLACE

Tanjung Simpang Mengayau—Sabah, Borneo

If you want to see a beautiful sunset without the crowds, visit the Tip of Borneo, a headland right at the very top of the largest island in Asia. Standing there, you have ocean views facing both east and west, which means you can actually see the sun rise and then set over the South China and Sulu Seas. It's a scenic place; think of a long white-sand beach with crystal-clear blue water, rocky cliffs, and vegetation galore. And while some may feel the drive to the tip is too long, the area's remoteness means less light pollution, so it's a perfect spot for stargazing at night, too.

MARVEL AT THE MOON IN MOAB

Arches National Park—Moab, Utah, United States

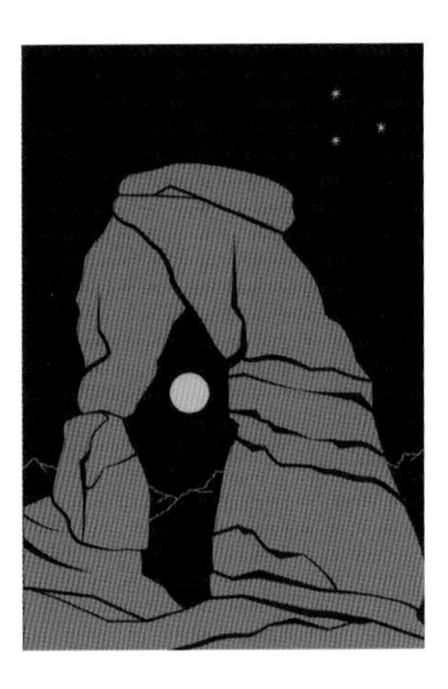

The United States has a huge supply of treasured national parks, and Arches National Park in eastern Utah is just one gem among many. Aside from its otherworldly red-rock landscapes, there's a lot going on in the skies above, too. First of all: The. Sunsets. Are. Incredible. And while moonless nights are better for stargazing, if you visit during a full moon (or, even better, a supermoon), you could glimpse our orbiting sky rock as it lines up inside the iconic Delicate Arch. Moon photographers meticulously plan to get their timing and positions right for this, so we won't pretend to know how to do it. But if you figure it out, we salute you!

Travel Tip: If you're going stargazing, preserve your night vision by using a red flashlight.

Shout-Out to Darksky.org: The International Dark Sky Association exists to protect the world's beautiful night skies through a global conservation program and is a leading figure in combating light pollution. It's also a great resource to find the best stargazing spots around the world, listing more than 130 certified Dark Sky Places on an interactive map.

PARTY WITH THE STARS IN LAS VEGAS

Death Valley National Park—Border of California and Nevada, United States

The International Dark Sky Association has put Death Valley in their Gold Tier, which means it has the highest possible darkness rating; it also happens to be the largest Dark Sky National Park in the world. So when someone invites you to a star party in Death Valley, you go. What's a star party, you ask? It's an event where amateur astronomers and casual star admirers get together to gaze up at the galaxies and nebulae and ponder the meaning of life. Sometimes there are laser-guided tours of the skies, and you might even be able to look through a Las Vegas Astronomical Society member's telescope if you ask nicely. You'd never guess that something so peaceful could exist *so* close to Sin City, but it does.

Good to Know: There are star parties year-round all over the United States and Canada, so there's a good chance you might also find one closer to home.

STARGAZE IN THE ATACAMA DESERT

Atacama Desert, Chile

Farther south lies another Dark Sky darling, the Atacama Desert in northern Chile. This sprawling plateau is the driest desert on Earth, and its clear skies and high altitude make it an ideal place for serious stargazers. You can visit one of the many observatories in the desert or simply sit on the balcony of your hotel and take in the wondrous star blanket from there. The Atacama Desert is called "Mars on Earth," and it truly does feel like being on another planet.

FIVE PLACES TO SEE THE NORTHERN LIGHTS

The aurora borealis is a spectacular display of glowing, dancing sky magic that everyone should witness once in their lifetime. If you want to see it, head north in the wintertime and be very patient, because you never know when it's going to appear. Here are just a few places you might spot it.

1. **SVALBARD, NORWAY**

   Svalbard has about ten weeks of round-the-clock darkness during winter, so up here you can actually see aurora borealis during the daytime, too.

2. **FAIRBANKS, ALASKA, UNITED STATES**

   Fairbanks's location makes it a sweet spot in Alaska (and one of the best in the world) for viewing the lights. It doesn't even have to be during the winter.

3. **LAPLAND, FINLAND**

   Want to ride a horse silently through snow-covered pines while you gaze up at the dancing sky? Do it in Lapland.

4. **REYKJAVÍK, ICELAND**

   There are so many places to see the Northern Lights in Iceland, but visit the Kvika foot bath near Grótta Lighthouse in the country's capital to keep your feet toasty while you wait for the sky to do its thing.

5. **SHETLAND ISLANDS, SCOTLAND**

   Visit Shetland for the rugged landscapes and curious traditions—not just to see the lights. If you're lucky, you might also spot the Mirrie Dancers, as they're known locally.

Meet Aurora's Cousin Steve: In the past few years, aurora chasers and citizen scientists have been documenting a new sky phenomenon that looks more like a vertical stripe of purplish green. It's called the Strong Thermal Emission Velocity Enhancement (STEVE). If you're out looking for Northern Lights, see if you can spot Steve, too.

PART THREE

THRILL SEEKER

Ever wondered what it is about scary—even terrifying—things that can make us feel so alive? Turns out there's a science to it: when we're faced with a perceived threat, our brains release mood-boosting chemicals like adrenaline, dopamine, and endorphins that kick us into gear and get our hearts beating rapidly.

Some people hate feeling like this, while others find it totally addictive. And it's no wonder people who fall into the latter camp are naturally drawn to travel, seeing the entire world as their playground to explore. There's no shortage of exhilarating experiences to be had up in the sky, under the sea, deep in the outdoors, or in the middle of a city—we know, we've looked! Venture through these pages, and hopefully you'll discover a new thrill or two that you'll want to chase around the globe.

But wait! Not to get all Mom on you, but before you plunge in, just remember that safety is also an essential part of any good adventure. Look after yourself, the people around you, and the environment—you can still take risks without being reckless. Okay, we're done. Now go forth and get out of that comfort zone!

UP IN THE AIR

Looking to get some perspective? Look no further. From reverse bungee jumping in Peru to riding a bike 150 feet in the air above forests in the Philippines, these adrenaline-pumping experiences will take you to new heights.

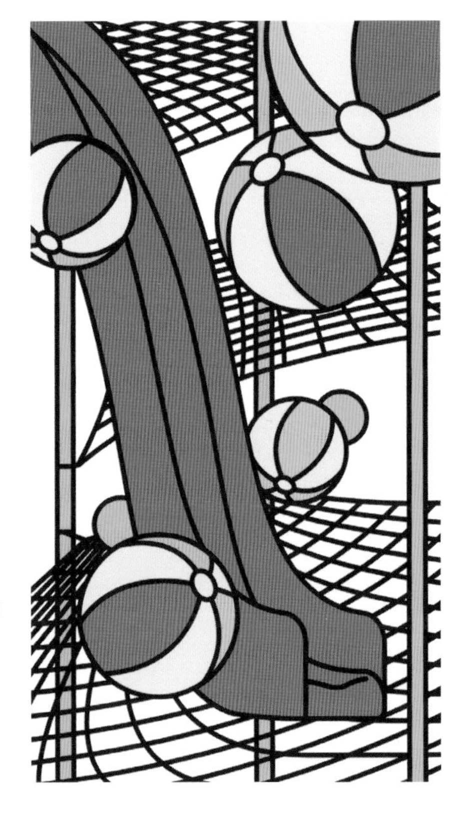

BOUNCE AROUND IN A SUSPENDED PLAYGROUND

AIRZONE—Singapore, Singapore

AIRZONE claims to be the world's first indoor atrium net playground, and it hangs inside City Square Mall, an eleven-story shopping mall in Singapore. Suspended between the second and sixth floors, the different levels are connected by maze tunnels and slides. Once you get over your fear of stepping out onto the nets and seeing just how far down the floor is (FYI, it's about 118 feet, or 36 meters), you can bounce around, swim through the floating ball pit, and play with giant inflatable orbs. Those who prefer to keep their feet on solid ground will be happy to know it's very entertaining to watch from the sidelines, too.

CLIMB YOUR HEART OUT AT A FOUR-LEVEL AERIAL PARK

SOAR Adventure Tower—Franklin, Tennessee, United States

Ready for a challenge? SOAR Adventure Tower has four levels of climbing, which start relatively low and easy but get progressively higher and harder—the top level sits 45 feet above the ground! There are more than 110 different climbing elements, like suspended bridges and swings, including some musical-themed ones (it is just 15 miles south of Nashville, after all). If you're looking for even more of a thrill, the park is open for twilight climbs so you can do it all in the dark.

RIDE ABOVE FORESTS ON A PEDAL-POWERED ZIP LINE

Chocolate Hills Adventure Park—Bohol, Philippines

You've never been on a bike ride like this before. On the island of Bohol in the Philippines, you can cycle 150 feet (45 meters) in the air on a pedal-powered zip line. Go it alone or grab your mate and make it a race. Either way, be sure to soak up the epic views of the surrounding Chocolate Hills, a stunning landscape covered with grassy mounds (they turn brown during the dry season, hence the name) that stretches on for miles. It's a truly surreal sight.

FIVE PICTURESQUE PLACES WHERE YOU CAN WALK AMONG THE TREETOPS

Whether you find them tranquil or terrifying, these sky-high canopy walks are sure to be ones to remember.

1. **CAPILANO SUSPENSION BRIDGE PARK—VANCOUVER, CANADA**
   A whole park dedicated to treetop adventures? Yes, please. Take the Cliffwalk 700 feet (213 meters) along cantilevered and suspended walkways through lush rain forest, or brave the 450-foot bridge suspended 230 feet (70 meters) over the Capilano River.

2. **CANOPY WALKWAY AT LEKKI CONSERVATION CENTRE—LAGOS, NIGERIA**
   At 1,315 feet (401 meters), this suspended swinging bridge is believed to be the longest in Africa. An epic view of the vegetation is guaranteed, but be on the lookout for crocodiles, monkeys, and a variety of birds, too.

3. **MULU CANOPY WALK—GUNUNG MULU NATIONAL PARK, MALAYSIA**
   One of the longest in the world, this treetop canopy walk extends for 1,378 feet (420 meters) and hangs 85 feet (25 meters) above the ground. Bird enthusiasts, come early for the chirpiest experience.

4. **REDWOODS TREEWALK—ROTORUA, NEW ZEALAND**
   Explore the forest of 118-year-old redwood trees from above on this network of twenty-eight suspension bridges and twenty-seven platforms. The whole walk is about 2,300 feet (700 meters) long. For an even more serene experience, come at night when the trees are lit up by lanterns.

5. **SELVATURA PARK—MONTEVERDE, COSTA RICA**
   Walk through the cloud forest on nearly two miles of treetop walkways and suspension bridges. You might even see a couple of monkeys!

SPOT ANIMALS FROM NEW HEIGHTS ON A BALLOON SAFARI

Masai Mara National Reserve—Narok County, Kenya

On their own, floating in a hot-air balloon and going on a safari are both pretty cool activities, but when you put them together? Now we're talking. Found in southwest Kenya, Masai Mara National Reserve is regarded as one of the best wildlife viewing spots in East Africa—and the most unique way to experience it is from the sky in a hot-air balloon safari. Come between August and November for a chance to witness the great wildebeest migration, when more than a million wildebeest (plus some zebras and gazelles) move up from Tanzania to the Mara.

SWING OVER THE "END OF THE WORLD"

Casa del Arbol—Baños, Ecuador

You know those extreme travel photos you see on the Internet and think, "That's gotta be fake!"? That's the kind of visual that awaits you at Casa del Arbol on the edge of Sangay National Park in Ecuador. Sitting atop a hill not too far from the town of Baños, there's a quaint tree house and lookout with views of lush green canyons and, on a clear day, Tungurahua Volcano (which is active, BTW). But the real attraction is the rather flimsy-looking swing that hangs from a tree branch and offers daring travelers the chance to soar out over the valley, seemingly at the edge of the world. Definitely take some snaps—just don't blame us if no one believes they're real.

FLOAT WEIGHTLESSLY ON A ZERO-GRAVITY FLIGHT

Zero-G—Various Locations, United States

A trip to space may not be in the cards for most of us, but, thankfully, you don't need to leave the stratosphere to experience the feeling of total weightlessness. Enter: the Zero-G Experience, where you can fly, float, and flip through the air like gravity doesn't even exist. It all takes place on a modified Boeing 727 aircraft: the plane performs parabolas, a maneuver that involves rising and diving in an arc shape, causing a zero-gravity effect. The sensation has been described as like floating in water—minus the water. Flights operate in various cities across the United States, and while the experience doesn't come cheap, it's still cheaper than going to space.

CLIMB WESTERN AUSTRALIA'S EXTREMELY TALL TREES

Gloucester National Park—Pemberton, Australia

Between the 1930s and 1950s in the southern corner of Western Australia, a number of fire lookouts were built at the top of a handful of very tall karri trees—a variety of eucalyptus that can live to more than 350 years old—and visitors can still climb a few of them today. Scale 174 feet (53 meters) up the Gloucester Tree in Gloucester National Park, or attempt the more recently pegged Dave Evans Bicentennial Tree, which will lead you to a 213-foot (65-meter) lookout, providing unrivaled 360-degree views of the forest.

BE FLUNG INTO THE SKY BY A GIANT SLINGSHOT

Action Valley—Cusco, Peru

If you've ever wished you could fly, your dream is about to come true. Not too far outside Cusco in the Peruvian Andes, there's a huge slingshot just waiting to launch you into the air like a firecracker. It's kind of like a reverse bungee jump, and it'll send you soaring more than 420 feet (128 meters) in only three seconds—certainly not an activity for the fainthearted. Before you commit, just remember the old saying: what goes up . . .

STAND ON THE WING OF A FLYING AIRPLANE

AeroSuperBatics—Gloucestershire, England

Walking out onto the wing of a plane while it's midflight sounds like an activity that should be reserved for people performing action-movie stunts, but it is in fact a very real thing you can do. The team of professional wingwalkers at AeroSuperBatics in the United Kingdom will coach you through the process step-by-step until you're standing on the wing of a classic 1940s biplane as it cruises through the sky. But that's only the beginning—if you want it to be. Extra-adventurous travelers can have the aircraft climb even higher, up to 600 feet (183 meters) and perform dips and dives at nearly 140 miles per hour. (We'll stick to the gentle cruising flight, thank you very much.)

WATCH YOUR FOOTING ON THIS DRAMATIC LOOKOUT OVER NORWAY'S FJORDS

Stegastein Lookout—Aurlandsfjord, Norway

Nothing but a single pane of glass stands between the edge of this viewing platform and the fjords below. Also called the Aurland Lookout, this wooden structure juts out 98 feet (30 meters) from the side of the mountain at a height of more than 2,000 feet (650 meters). Designed by architects Tommie Wilhelmsen and Todd Saunders, it might look disconcerting, but it's totally safe and a great spot for panoramic photos of the fjords and the surrounding mountains. You'll find it about a four-and-a-half-hour drive from Oslo on the National Tourist Road, which travels from Aurland to Laerdal.

Legs Feeling Wobbly Just Reading About These Experiences? You're Not Alone: The term for an extreme fear of heights is "acrophobia," and it's thought to be one of the most common phobias. So you're in good company!

FACE YOUR FEARS ON AN EXTREMELY HIGH ROPES COURSE

The Gravityz—George Town, Malaysia

Scared of heights? The Gravityz might actually help change that. Set 784 feet (239 meters) high on the side of a skyscraper in the middle of George Town, Penang, this obstacle course is designed to help people build confidence by literally stepping out of their comfort zone. There are six challenge obstacles in total, which get more intense as you move through the course. Start with the Confidence Path, a basic walkway that hugs the building, before progressing to the Great Bridge, a narrow overpass with sheer drops on either side, and finally the G Rocky, which requires you to lie on your back on a small platform. If you can muster the courage, do try to look down—the view is pretty spectacular.

UNDERWATER ADVENTURES

Seventy-one percent of the Earth is covered in water, which means there are endless adventures to be had viewing the natural world below the surface. Dive into the next pages for stunning snorkels, thrilling swims, and peaceful encounters with apex predators.

CAGE DIVE WITH SALTWATER CROCODILES

Crocosaurus Cove—Darwin, Australia

You can't get friendly with a crocodile—unless you're in the Cage of Death. This spine-tingling experience at Crocosaurus Cove in Australia's Northern Territory lets you come snout-to-snout with a five-meter-long saltwater crocodile, all from the safety of a perspex cage. Only crocs that are deemed "in the mood" are invited to swim with guests, and in exchange they get tasty nonhuman meat snacks. After you've had your croc meet, you can swim (sort of) with kid crocodiles in their swimming pool and get to know other scaly critters in their big reptile display.

Fun Fact: Saltwater crocodiles look about the same today as they did two hundred million years ago. So this is the next best thing to swimming with dinosaurs!

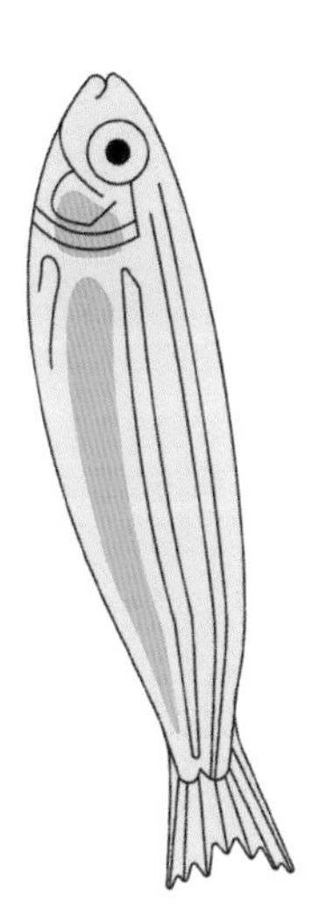

SNORKEL WITH A MILLION SARDINES

Panagsama Beach—Moalboal, Philippines

If swimming in the chaotic energy of millions of sardines sounds fun to you, you might want to think about a visit to the famous Moalboal Sardine Run on Cebu island in the Philippines. The sardines hang out about twenty meters from the shore, and if you wade out from Panagsama Beach, there's a good chance you'll be able to see schools of them swimming back and forth in search of tasty plankton. If you need to, you can rent snorkels, life jackets, and other equipment from the Savedra Dive Center at the beach. But swimming alongside the tiny, oily fish on your first-ever sardine run is completely free of charge.

Travel Tip: Head to Cebu in the dry season (November–May) for better snorkeling visibility.

SNORKEL ALONG THE "UNDERWATER GRAND CANYON"

Andros Island, Bahamas

The Bahamas boasts an extensive underwater cave system, also known as the mysterious Blue Holes, and Andros Island is where serious divers head to explore astonishing passageways and caverns filled with dripstones and stalagmites. If you want to experience the blue holes without entering them (looking at you, snorkelers!), head to south Andros to float and swim in the waters above the caves' entrances. Even better, snorkel down "the crack" at Tiamo: a long limestone crevice that will make you feel like you're swimming over a tiny Grand Canyon filled with tropical fish.

FRESHWATER DIVE IN SHARK-FREE LAKE MALAWI

Lake Malawi—Mozambique, Tanzania, and Malawi

Lake Malawi is known as the "inland sea" of Malawi, and it's one of the best freshwater diving (and snorkeling!) spots in the world. The lake is about 365 miles long and runs through Mozambique and Tanzania as well; different parts of the lake are better to visit depending on the type of experience you're looking for. Dive among granite boulders and colorful *mbuna* (rockfish) in the UNESCO World Heritage–listed Lake Malawi National Park, or explore the crystal waters of the lake from Manda Wilderness in Mozambique and see if you can spot an otter. The best part? It's a lake, so you don't have to worry about sharks here.

GO ON AN UNDERWATER SHARK SAFARI IN FIJI

Shark Reef Marine Reserve—Beqa Island, Fiji

If sharks are what you're looking for, visit Beqa Island in Fiji for what some call the best shark dive in the world. Beqa Adventure Divers worked with the Fiji government and the traditional owners of the reef to create a marine conservation project (and what is now Fiji's first National Marine Park) that protects eight different species of shark and hundreds of other species of marine life. Their shark dive takes you into the lagoon to swim among the naturally roaming creatures, including different reef sharks, bull sharks, and even the occasional tiger. It's really a shark lover's dream.

SWIM THROUGH THIS SUBMERGED SOUTH AFRICAN GARDEN

Marico Oog—Lichtenburg, South Africa

To get up close and personal with some incredible aquatic plant life, visit Marico Oog, a natural spring and series of connected freshwater pools about two hours from Johannesburg. Once you dive into the warm water, you'll find yourself in an enchanted underwater forest, swimming through a corridor of reeds as the sun shines through a ceiling of water lilies. It's truly magical. It's also an amazing spot for night dives, where you can actually stargaze from underwater.

Good to Know: Marico Oog runs across a series of private farming properties. If you do visit, be respectful of the fences blocking off certain areas.

CHECK OUT LOBSTER AVENUE IN GOA

St. George's Island—Goa, India

Goa is known for its beautiful beaches and scuba-diving spots, and with their shallow, still waters and beautiful reefs, the islands in the region are perfect for divers of all levels. If you want to see colorful marine life and coral, Lobster Avenue is a good place to start. It's a dive site on the south side of St. George's Island where you'll find communities of lobsters living among the rocks and coral reefs. Aside from exotic crustaceans, there are also colorful fish and turtles living down here. And at ten meters deep, it's a fairly shallow dive, so it's suitable for beginners.

EXPLORE AN UNDERWATER ARCHAEOLOGICAL PARK

Old Caesarea Diving Center—Caesarea, Israel

Caesaera is a national park, restored harbor, and archaeological site on the Mediterranean coast, about two hours' drive from Tel Aviv. History lovers will want to visit the Caesarea Underwater Archeological Park, the world's first, to explore the ruins of the ancient harbor built by Herod, who founded Caesarea in honor of Julius Caesar. You can snorkel and dive through the sunken port and get a feel for harbor life in 21 BCE, and there are shipwrecks to explore, too. When you're done with the underworld, take a stroll along Caesarea Aqueduct Beach to watch the sunset with a backdrop of even more ancient Roman ruins.

QUINTANA ROO, MEXICO

Swimming with Whale Sharks in Mexico

Did you know that whale sharks are not actually sharks but gigantic filter-feeding fish? So when you say you went swimming with whale sharks, it sounds way more terrifying than it actually is. Mexico's Yucatán Peninsula is famous for its whale-shark swimming experiences, but it's worth looking for an operator with an ethical practice, because crowds of tourists that come to see these gentle giants can be damaging to their well-being. The smaller, quieter Holbox Island is a great place to look for an eco-conscious whale-shark adventure.

The World Wildlife Fund has worked with local fishermen on a set of guidelines for operators and tourists who want to swim with whale sharks. Among other things, the rules include keeping the boat a minimum distance from the sharks, not touching the animals, and not using flash photography.

DIVE IN THIS EPIC ABYSS IN BRAZIL

Abismo Anhumas—Bonito, Brazil

Adventurers who want to combine rappelling, diving, and climbing should visit the epic 236-foot-deep Anhumas Abyss in the Brazilian state Mato Grosso do Sul. You're lowered into the cave (claustrophobes beware; the first half of the descent is through an extremely narrow opening), at the bottom of which you'll find a crystalline blue lake. There, divers continue their descent exploring the depths of the lake, discovering stalagmites and the fossilized bones of animals who have sadly fallen into the abyss. If you're not a certified diver, you can also snorkel or just float around the cave on a small boat. Eventually, you'll make a slow climb back up the walls of the cave to continue your life aboveground.

SLEEP UNDERWATER AT THESE TWO UNIQUE HOTELS

Spend the night with the fishes, have a very watery wedding, and even order pizza under the sea. It's all possible!

1. **JULES' UNDERSEA LODGE—KEY LARGO, FLORIDA, UNITED STATES**
   The lodge is a converted research lab in a mangrove lagoon at Key Largo Undersea Park. Certified divers can dive down to this famous underwater hotel to spend the night—they even offer underwater pizza delivery. If you're feeling extra romantic, you can also get married underwater first, with your wedding officiated by a Justice of the Pisces.

2. **HOTEL UTTER INN—VÄSTERÅS, SWEDEN**
   This artist-designed cozy underwater hotel room sleeps only two people at a time and sits a few meters beneath a red floating jetty on Lake Mälaren. You can book to stay here from April to October; in winter the lake freezes over!

HIKES AND TREKS

The Andes, the Swiss Alps, the Himalayas, literally every national park in the United States and Canada, pretty much the entire country of New Zealand—we humans are really spoiled for choice when it comes to extraordinary hiking destinations. It'd be a near-impossible feat to cover every trail worth trekking, but on the next pages you'll find some interesting hikes to get you started.

HIKE TO A FAIRY TALE–LIKE GLACIAL LAKE

Laguna 69—Huaraz, Peru

When it comes to must-do hikes in Peru, the Inca Trail gets most of the airtime, but there's another, lesser-known one that should be on your radar. Laguna 69 is a piercingly blue, out-of-this-world-beautiful glacial lake in Huascarán National Park that you can reach only via a strenuous hike. The trail passes by waterfalls and mountains and is a little more than four miles (one way) with a 2,700-foot elevation gain—but it's the altitude, not the distance that makes this journey really tough. If you can muster the energy, though, the view at the end is 100 percent worth it.

TEETER ACROSS THE LONGEST HANGING PEDESTRIAN BRIDGE

Europa Trail—Zermatt, Switzerland

The Swiss Alps are no stranger to breathtaking hikes, and the Europa Trail is widely thought to be one of the most beautiful. Best tackled over two days, this challenging trail runs for 21 miles (34 kilometers) from Grächen and Zermatt and offers fantastic mountain views, including a peek at the famous Matterhorn. Hikers will also have to tackle a very long suspension bridge (the longest in the world, according to local authorities), which stretches on for 1,620 feet (494 meters) and hangs almost 280 feet (85 meters) high above the Grabengufer ravine. Crossing the bridge takes about ten minutes, so you'll have plenty of time to mildly freak out about how bloody high up you are while also trying to soak up the scenery.

The Golden Rule of Exploring the Outdoors–Leave No Trace: The Seven Principles of Leave No Trace encourage us to leave the environment as we found it, to respect wildlife, to adopt a pack-in and pack-out trash policy, to travel and camp on durable surfaces, and more. The guidelines were created by the Leave No Trace Center for Outdoor Ethics as a framework for minimizing human impact on nature while visiting the outdoors. To memorize the seven principles, which we probably all should, visit lnt.org/why/7-principles.

CHANNEL YOUR INNER HOBBIT ON A JOURNEY TO MOUNT DOOM

Mount Ngauruhoe—Ruapehu, New Zealand

Fans of the *Lord of the Rings* movies will recognize Mount Ngauruhoe as the fiery Mount Doom. But here's some good news: you don't have to brave Mordor in order to visit this impressive volcano, which stands 7,500 feet (2,287 meters) tall in Tongariro National Park. For a memorable journey of your own, hike the Tongariro Alpine Crossing along the saddle between Mount Tongariro and Mount Ngauruhoe and take a side trip up to the crater rim (look out for steam vents). It's a solid one-day hike that will reward you with epic scenes of lakes and Mars-like landscapes. The volcano last erupted in 1975, but it is still active, so be careful and go at your own risk.

CHASE WATERFALLS IN THE DOMINICAN REPUBLIC

27 Charcos de Damajagua—Puerto Plata, Dominican Republic

Okay, maybe "jump down waterfalls" would be more accurate. Not too far from Puerto Plata, the Damajagua River boasts twenty-seven waterfalls, and you can experience them all on a one-day adventure. It starts with a hike up through dense forest to the top of a rocky hill. Then the real fun begins as you make your way back down the river, sliding and jumping down the falls into stunning blue pools. (Some drops are up to thirty feet!) There are local guides to help you navigate the journey; if you don't have the energy for all twenty-seven, shorter hikes will take you to the seventh or twelfth fall.

SIX ESSENTIAL ITEMS TO TAKE ON EVERY HIKE

Hiking is a simple pleasure, but for maximum safety, relaxation, and enjoyment, you need to be prepared. This is by no means a complete list—longer or overnight hikes require more planning and equipment—but here are a few basic day-hike items that'll help you get the most out of your time on the trail.

1. **A map:** If you're headed to a national park, stop by the visitor center to pick up a trail map or hiking guide. For remote wilderness and backcountry hiking, bring a GPS system, topographical map, and a compass (and make sure you know how to use them).

2. **Proper hiking shoes:** There's nothing worse than feeling a blister forming on your heel halfway through a hike. But this is not just a matter of comfort; it's a safety thing, too. It's worth investing in good-quality shoes with proper tread, ankle support, and water-resistant fabric.

3. **Drinking water:** Ensuring you have adequate water to stay hydrated is arguably the most important tip—it's thirsty work out there! Fill up reusable water bottles or one of those fancy water bladders with a straw, and pack some water purification tablets, just in case.

4. **Sun protection:** Grab your sunnies, a hat, and sunscreen to block those harsh UV rays—even in winter.

5. **Snacks:** Granola bars, trail mix, nuts, and fruit are great sustenance foods. Hiking can be a solid workout, so bring more than you think you'll need.

6. **Extra clothing:** You might work up a sweat on the trail, but your body will cool down once you stop moving. Pack a couple of extra layers that you can throw on during your rest breaks. A compact waterproof jacket is always a good idea, too.

WATCH THE SUNRISE ATOP A SACRED MOUNTAIN

Mount Sinai—Sinai Peninsula, Egypt

Located in the east of Egypt, Mount Sinai (or Jabal Mousa as it's known locally) is an important religious site among Jews, Christians, and Muslims; it's supposedly where Moses received the Ten Commandments from God. The mountain's peak sits at 7,497 feet (2,285 meters), but the most well-trodden trails start 300 meters below the summit. The trek is popular among pilgrims and hikers (who should be accompanied by a local Bedouin guide) and most often attempted at night so as to reach the summit at sunrise for the most breathtaking views of the surrounding mountains and valleys below.

HIKE ACROSS JORDAN'S COUNTRYSIDE

Jordan Trail—Jordan

The Jordan countryside is connected top to bottom by a series of over 420 miles (675 kilometers) of trails, running from the northern town of Umm Qais to the southern city of Aqaba. The Jordan Trail passes through seventy-five towns and villages and traverses incredible landscapes: the green hills in the north, the cliffs of Jordan Rift Valley, the otherworldly scenes of Wadi Rum desert, and everything in between. Unless you're a serious (like, really serious) hiker, chances are you'll probably just want to tackle one section, at least at first—the entire route would take around forty days to complete. If you're not sure where to start, the Dana to Petra region is pretty spectacular.

SPOT OTTERS ON SOUTH AFRICA'S MOST ICONIC COASTAL TRAIL

Otter Trail—Garden Route, South Africa

Even just taking a single step on this scenic trail isn't easy; bookings need to be made up to a year in advance, and only twelve people can commence the hike per day. But its popularity is well earned. The five-day hike in Garden Route National Park traverses 28 miles (45 kilometers) of coastlines, cliff tops, forests, and rivers, with breathtaking scenery all around. It's a fairly strenuous hike; you'll need to contend with river crossings (prepare to get wet!) and carry enough food for the entire journey, but the vistas will make it all worthwhile. Wildlife enthusiasts, keep your eyes peeled: you might be able to spot dolphins, monkeys, and clawless otters (the ones that gave the hike its name) along the way.

Hiking Tip: Before you set out on any hike, always tell a friend or family member exactly where you're going and when you'll be back. If anything goes wrong **touch wood** and they don't hear from you, they'll know to get help.

PRETEND YOU'RE ON ANOTHER PLANET AS YOU HIKE AMONGST HOODOOS

Bryce Canyon National Park—Garfield County, Utah, United States

You won't find more hoodoos (eroded spires of rock) anywhere in the world than in Utah's Bryce Canyon National Park; everywhere you look, irregular columns of red (and orange and pink) rocks jut into the sky, creating a fairy tale–like atmosphere. After you've admired it from above, take the Peek-a-Boo loop trail down into Bryce Amphitheater for a glorious 5.2-mile (8.4-kilometer) hike that'll have you exclaiming, "Oooh, cool!" and "Wow!" around every corner. The secret's out, though: the park gets more than two million visitors each year. To beat the crowds, come during the cooler months when the trails are much quieter. Plus, the hoodoos look even more magical with a light dusting of white snow.

TREK PAST TURQUOISE LAKES TO A TEA HOUSE

Lake Louise, Banff National Park—Banff, Alberta, Canada

Lake Louise is all about the drama. You've probably seen pics of it—an idyllic turquoise lake surrounded by snowcapped mountains—and it's even more beautiful IRL. While it might be tempting to sit and stare longingly into the water all day, you'd be missing out if you did: there's a hike with tea and pie at the end, people! A well-trodden path leads you about 2.2 miles (3.5 kilometers) around Lake Louise and up past Lake Agnes, before arriving at Lake Agnes Tea House, a historic café that has been serving up tea to hikers since 1905. Take a seat on the balcony and enjoy the scenery—you've earned it.

MUSTER THE COURAGE TO HIKE CHINA'S MOST TREACHEROUS TRAIL

Hua Shan—Huayin, China

Mount Hua (Hua Shan) in China's Shaanxi Province is a sacred Taoist site, and the climb up is often referred to as one of the most dangerous hikes in the world. There are five granite domes, the tallest of which is South Peak at 7,087 feet (2,160 kilometers). This dome is where you'll find the plank walk: a series of narrow wooden planks bolted into the side of the mountain. Hikers are harnessed in, but even with some support the sheer drop is pretty unnerving. If you dare brave it, be very careful, but make sure you pause for a second to appreciate the dreamlike scenery. And remember: it's a holy mountain, so please be respectful.

What's with the Piles of Rocks Everywhere? If you've been on a remote-ish hike before, you may have spotted piles of carefully stacked rocks. These are called cairns, and they're typically intended as markers to help guide hikers along a trail. In some parks these are maintained by rangers, but that's not always the case; it's best to read up on the trail's navigation before following them (and take a map!). So next time you're hiking and you come across a cairn, please don't tamper with it. Also, don't build your own—we know they look pretty, but moving rocks can disturb the soil, vegetation, and creatures beneath.

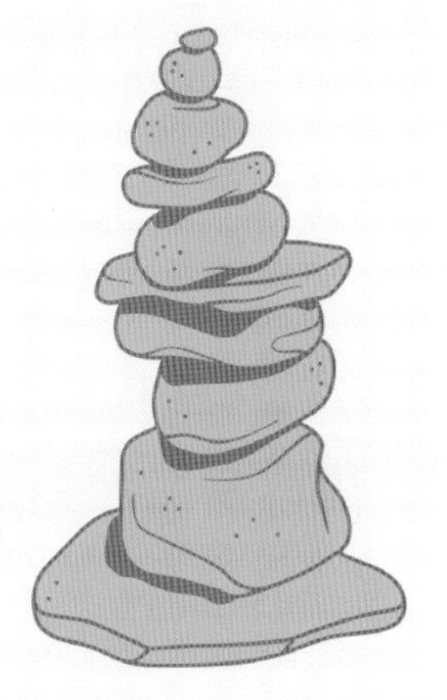

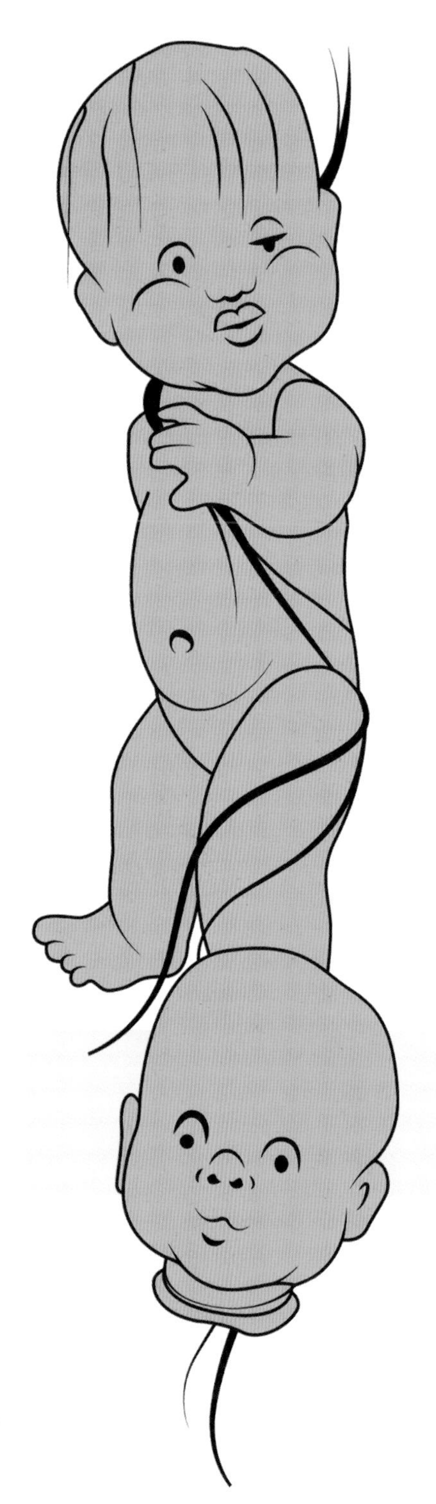

SPOOKY SPOTS

If you're an advocate for Halloween-all-year, step this way for ancient mysteries, ghosts galore, creepy clowns, and islands covered in snakes and haunted dolls.

FLOAT PAST THIS ISLAND COVERED IN HANGING DOLLS

The Island of the Dolls—Mexico City, Mexico

Imagine you're drifting down a canal on a boat and look to your left to see hundreds of plastic baby dolls hanging from the trees of an empty island. Some of the dolls have no eyes; others have no heads. Welcome to La Isla de Las Muñecas in Xochimilco—the Island of the Dolls. As is most often the case with spooky places, there's a sad story behind it: supposedly, the island's caretaker said he saw a young girl drown in the canal, and when he later found a floating doll nearby he strung it up in the trees (along with many other dolls over the years) to please the girl's spirit. Eventually, the man also drowned, and the dolls now hang there with no caretaker.

Psst! Sad and creepy dolls aside, the Xochimilco canal is a cool place to visit. Read more about it on page 24.

VISIT AN ANCIENT CEMETERY IN RUSSIA

City of the Dead—Dargavs, Russia

In Russia's Republic of North Ossetia, Alana, not far from the village of Dargavs, there's a burial site that dates back to the sixteenth century known as the City of the Dead. It's set on farmland surrounded by picturesque mountains, and ninety-nine hut-like stone crypts dot the hillside in their own tiny necropolis. More than ten thousand bodies are said to be buried in this ancient Ossetian grave city, although nobody knows exactly who they are. It's spooky, but it's also a fascinating place for lovers of history and natural beauty.

THREE PLACES WHERE YOU MAY OR MAY NOT FIND GHOST CATS

Human ghosts are spooky, but ghost cats? Kinda fun. Here are three places that are said to be haunted by phantom kitties.

1. **FARNAM MANSION—ONEIDA, NEW YORK, UNITED STATES**
   This famously haunted mansion is supposedly full of spirits, including that of at least one feline.

2. **HELL-FIRE CLUB—DUBLIN, IRELAND**
   One of Ireland's most haunted buildings. Around here you might encounter the Killakee cat, a giant black ghost cat with fiery red eyes.

3. **STANLEY HOTEL—ESTES PARK, COLORADO, UNITED STATES**
   This is the place that inspired Stephen King's *The Shining*. Staff claim that a gray cat haunts the tunnel system beneath the hotel.

VISIT THIS "HOUSE OF TEARS" IN MEXICO

La Casa Figueroa—Taxco, Mexico

If a home has been named the House of Tears by locals, you know something terrible has happened there. Such is the case with La Casa Figueroa in Taxco, a mountain town south of Mexico City. The house was built in the eighteenth century for a count by people of the Tlahuica tribe, and the tribe members were treated so badly that it's believed that since then the house has been cursed. When artist Fidel Figueroa bought it in 1943, he discovered secret rooms, tunnels, and vaults throughout, and he decided to open the residence up to the public rather than live there. It's now a museum full of curious artifacts, traditional art, very disturbing vibes, and most definitely ghosts.

SPOT THE MYSTERIOUS MISSOURI SPOOKLIGHT

Hornet Spooklight—Near Hornet, Missouri, United States

The Hornet Spooklight (or Ozark Spooklight) is a light phenomenon that has been seen on the Missouri-Oklahoma border a few miles out from the small town of Hornet, described by some as a glowing orb about the size of a basketball. It's been reported by various people for more than a hundred years; still, nobody knows WTF it is. There's also a UFO hot spot around the same area called Marley Woods where a variety of paranormal and other strange sightings have occurred, including stories of the Spooklight. It's not clear exactly where Marley Woods is, but it's probably somewhere you don't want to get lost in late at night.

SLEEP (OR DON'T) AT THIS CLOWN MOTEL

The Clown Motel—Tonopah, Nevada, United States

It's a clown-themed motel next to a cemetery. Seriously. The Clown Motel in Tonopah is home to a collection of two thousand clowns, and every room has its own spooky backstory. Did we mention it's right next to a cemetery? It's called America's Scariest Motel, and rooms start at seventy dollars a night. But if you can't bear the idea of an overnight stay (and we don't blame you!), it's still a neat and spooky place to visit in the safety of daylight.

LOOK FOR PRISONER GHOSTS IN LATVIA

Karosta Prison—Liepaja, Latvia

"Welcome to prison!" says the website of this former military base built by the czar of Russia on the west coast of Latvia. Karosta was built in the early twentieth century as a garrison prison to punish naval officers and is apparently extremely haunted. You can take a guided tour of the prison, and they have other interactive experiences, including an "Escape from the USSR" spy game, an Escape Room, and an "extreme" overnight stay where you're treated like an actual prisoner for the night. If that sounds like too much, you can spend a night in their hotel, sleeping in a prison cell but without all the nasty punishments.

DON'T ACTUALLY VISIT THIS SNAKE ISLAND IN BRAZIL

Ilha de Queimada Grande—São Paulo, Brazil

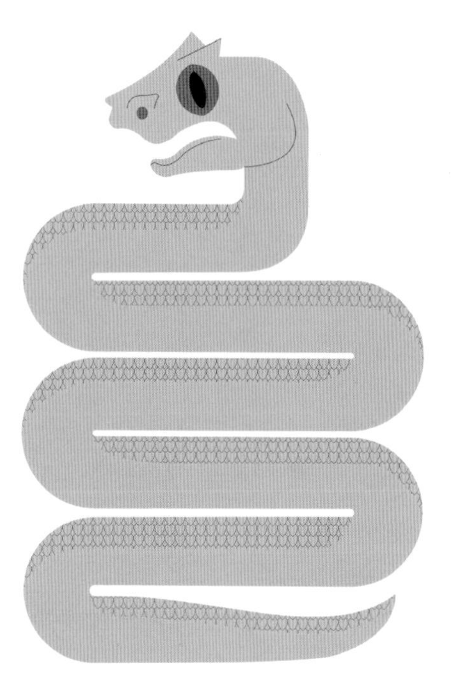

Here's a spooky place to know about but never actually visit: Snake Island in Brazil. Located off the coast of the state of São Paulo, the island's only full-time residents are a few thousand golden lancehead vipers, a.k.a. some of the deadliest snakes in the world. The Brazilian government won't actually let you visit the snake-infested island unless you're going on official research (and with a doctor present), so don't get any ideas. There are so many sneks living here and their venom is *so* incredibly toxic that you wouldn't last long if you showed up alone.

HOLD YOUR BREATH AT CRYBABY LANE

Crybaby Lane—Raleigh, North Carolina, United States

Crybaby Lane is a haunted (or at a minimum creepy) grassy area off Bilyeu Street in Raleigh. Legend has it that it's the location of a burned-down orphanage, and that if you stay there long enough you can smell the burning wood and hear the cries and wails of children. The story has been debunked—it's not the former site of an orphanage, although there was one nearby—but that doesn't make this place any less spooky. That said, it's literally a patch of grass and some abandoned houses, so don't plan your whole vacation around it.

VISIT A DEAD GLACIER

Okjökull—Ok, Iceland

Here's something not only spooky but downright terrifying you can visit: Iceland's first dead glacier, which after years of melting was finally declassified in 2014. Okjökull lived happily for centuries on top of the volcano Ok in western Iceland, but thanks to climate change has now been lost forever; in its place is a memorial plaque. It was the first of Iceland's four hundred glaciers to melt, and the rest are expected to follow in the next two hundred years. So if you want to be spooked and also reminded that climate change is real and happening all around us, next time you're in Iceland go and spend some time at the grave of Okjökull, the ghost glacier.

DAREDEVIL SPORTS

Buckle up and take a deep breath. From go-karting on city streets to volcano boarding to climbing a canyon wall of ice, these extreme sports are guaranteed to get your heart racing.

DRIVE GO-KARTS THROUGH THE STREETS OF TOKYO

Street Kart—Tokyo, Japan

It might not be Mario Kart–themed anymore, thanks to Nintendo's successful copyright lawsuit, but seeing the sights of Tokyo while you zoom through the streets on a go-kart is still bucket-list worthy, if you ask us. Street Kart (formerly MariCar) lets you do just that; all you need is a valid Japanese or international driver's license and the courage to get behind the wheel. There are a couple of different courses to choose from, and you can still dress up in costume—just not as Luigi or Bowser. And while it's supposed to be fun, remember it's not a video game: it's real life and they're real roads, so obey traffic laws and stay safe.

PLAY PAC-MAN IRL

Universal Games—Madrid, Spain

How's this for retro: at Universal Games activity center on the outskirts of Madrid, you can put on a Pac-Man costume and run through a big blue maze, collecting balls while people dressed as ghosts chase you around—just like in the '80s video game. There are more than ten levels of difficulty to complete, too, so you can play it over and over. Once you've nabbed a high score, there are tons of other sports to try your hand at, like paintball, a medieval challenge, and zombie war games.

SANDBOARD DOWN THE SIDE OF AN ACTIVE VOLCANO

Cerro Negro—Leon, Nicaragua

There's more than one reason Nicaragua's youngest volcano, Cerro Negro, will get your blood pumping. For starters, it's very active; it's erupted eighteen times in the past hundred years (the last recorded eruption was in August 1999). Second, it's a steep hour-long climb to get to the rim—but going back down is where it gets more interesting. The path is the most leisurely route; however, for those seeking maximum thrills, your best bet is to (a) make a run for it or (b) try your luck sand-surfing down on a wooden board. It's worth noting once more: there's no 100 percent safe way to visit a volcano. But risks aside, this black-sand beauty lures daring travelers from across the globe.

SEVEN WEIRD SPORTS AROUND THE WORLD YOU PROBABLY HAVEN'T HEARD OF

1. **BOG SNORKELING—POWYS, WALES**
   A swimming competition in which contestants don flippers, goggles, and a snorkel and race through a 197-foot (60-meter) muddy peat bog. The idea for the sport is believed to have originated in a bar (makes sense), and the world championship is held every August.

2. **CHEESE ROLLING—GLOUCESTERSHIRE, ENGLAND**
   Set on the *very* steep slope of Cooper's Hill, Gloucestershire, this annual race sees gutsy competitors run, tumble, and roll down the hill, trying to catch a rather large wheel of cheese. It's painfully funny to watch.

3. **COMPETITIVE SLAPPING—SIBERIA, RUSSIA**
   Essentially just a good ol'-fashioned slap-off, where two men take turns hitting each other in the face until one person eventually bows out. The Russian Male Slapping Championship usually takes place as part of the Siberian Power Show.

4. **TEJO—VARIOUS LOCATIONS IN COLOMBIA**
   This game, which can be played individually or in teams, involves throwing metal disks at a bunch of explosives—often with a beer in hand. It's an adaptation of an old Indigenous game, and in 2000 it was designated as the official national sport of Colombia.

5. **FINGERHAKELN (FINGER WRESTLING)—VARIOUS LOCATIONS IN GERMANY AND AUSTRIA**
   Also called "finger pulling," much like traditional wrestling this sport matches competitors by age and weight. But instead of entering a ring, they sit on opposite sides of a table and, using only one finger, pull as hard as they can on a small loop of leather. Whoever manages to pull their opponent across the table wins.

6. **TRUGO—MELBOURNE, AUSTRALIA**
   Best described as a combo of lawn bowls, croquet, and golf, this niche sport involves using a wooden mallet to hit rubber rings down a grass court and through a goal the width of a railway track. It was invented at railway workshops in Melbourne back in the '20s and has an aging player base—the current world champion is in his nineties.

7. **SEPAK TAKRAW (KICK VOLLEYBALL)—VARIOUS COUNTRIES IN SOUTHEAST ASIA**
   Popular in countries like Thailand, Indonesia, Myanmar, and Malaysia, Sepak Takraw is just like volleyball—except instead of using your hands to hit the ball, you have to use your feet. As you can imagine, it requires some serious athletic ability!

GO INDOOR SKIING IN THE MIDDLE OF THE DESERT

Ski Dubai—Dubai, United Arab Emirates

Is there anything *cooler* (sorry) than going skiing in the desert? We think not. Set inside a mall in central Dubai, you can take lessons from the resort's professional instructors, or go right ahead and hit the slopes on skis or a snowboard. If you're more interested in the AC than the physical exertion (no judgment), chill out in the snow park or make friends with the resident king and gentoo penguins. Oh, and remember to bring your winter gear—the temperature inside hovers around 30 degrees Fahrenheit (-1 degree Celsius).

PRETEND YOU'RE INDIANA JONES AT A JUNGLE ADVENTURE PARK

XPLOR Park—Playa del Carmen, Mexico

From zip-lining to jungle off-roading to underground rafting, when it comes to extreme sports, this adventure park has it all. Start up high on the zip-line circuits, which take you over the jungle and splashing through a cenote, and then jump in an amphibious vehicle and drive across hanging bridges and inside caverns. Here you can also hike, swim, and paddle through the underground caves, admiring the stalactite formations as you go. To take the thrills up a notch, visit at night for the Xplor Fuego experience. Indiana Jones would be proud.

GO WHITEWATER RAFTING IN THE MIDDLE OF THE CITY

Riversport—Oklahoma City, Oklahoma, United States

You don't even need to leave the city to experience the best outdoor activities; Riversport Adventures is an outdoor recreation park in the heart of Oklahoma City. The Riversport rapids offer as authentic a white water–rafting experience as you can get (apart from, you know, natural rapids). There are varying levels of intensity available, from Class II to IV, and if you're feeling daring you can take to the water in a kayak or tube instead. When you're ready to dry off, there's a 74-foot tube slide and BMX course with your name on it.

PRETEND TO BE A LUMBERJACK FOR A DAY

Wild Axe Park—Nova Scotia, Canada

Don your finest flannelette shirt and limber up for what might just be the most Canadian sporting activity ever. Run by Darren Hudson, a world champion lumberjack, the Lumberjack AXEperience will teach you all the classic moves: log rolling, tree climbing, cross-cut sawing, bow sawing, and axe throwing. You know the ones. The park is located in a rather serene setting on Nova Scotia's picturesque Barrington River, among plenty of towering pines—just don't go chopping them all down with your newfound lumberjack skills, okay?

RACE A LUGE THROUGH A REDWOOD FOREST

Skyline Luge—Rotorua, New Zealand

You know luge, that sport in the Winter Olympics in which athletes lie on their backs and race down icy courses at speeds of up to ninety miles per hour? A (much safer) version of the sport was developed in New Zealand; riders climb aboard a luge cart (kind of like a wheelless go-kart) and slide down tracks, with the help of gravity. Skyline Luge now operates in Canada, Singapore, and South Korea, but if you can make the journey, check out the original location in Rotorua, New Zealand, a spectacular volcanic area that's home to redwood forests.

CLIMB UP A FANTASY-LIKE WALL OF ICE

Ouray Ice Park—Ouray, Colorado, United States

You've heard of rock climbing, but what about ice climbing? Ouray Ice Park is an impressive man-made outdoor park in the Uncompahgre Gorge, just near the cute ski town of Ouray in southwest Colorado. Every year in the lead-up to winter, "ice farmers" spray the canyon walls with water (overflow from the City of Ouray), which freezes and builds up over time, creating imposing ice walls that look like they belong in an episode of *Game of Thrones*. In order to climb, you'll need the proper gear, like crampons and a helmet, which you can rent from vendors in town. The park is open seasonally, usually from mid- to late December until mid-March, and it's free to enter.

MOUNTAIN BIKE DOWN BOLIVIA'S MOST DANGEROUS ROAD

North Yungas Road—La Paz, Bolivia

Nicknamed "Death Road" (we'll let you guess why), North Yungas Road is a notoriously treacherous route that runs from Bolivia's capital, La Paz, to the forested Yungas region below. Cut into the side of the mountain, it's a windy descent that lasts about 12,000 feet (3,650 meters), and the road itself measures less than 12 feet (3.5 meters) wide, with a terrifying sheer drop down into the abyss. Oh, and did we mention it's a *two-way* road? None of that scares off the adrenaline junkies who come in hordes to cycle down the path. If you're up for the extreme challenge, you can organize a tour through most hostels in La Paz—but be aware of the risks.

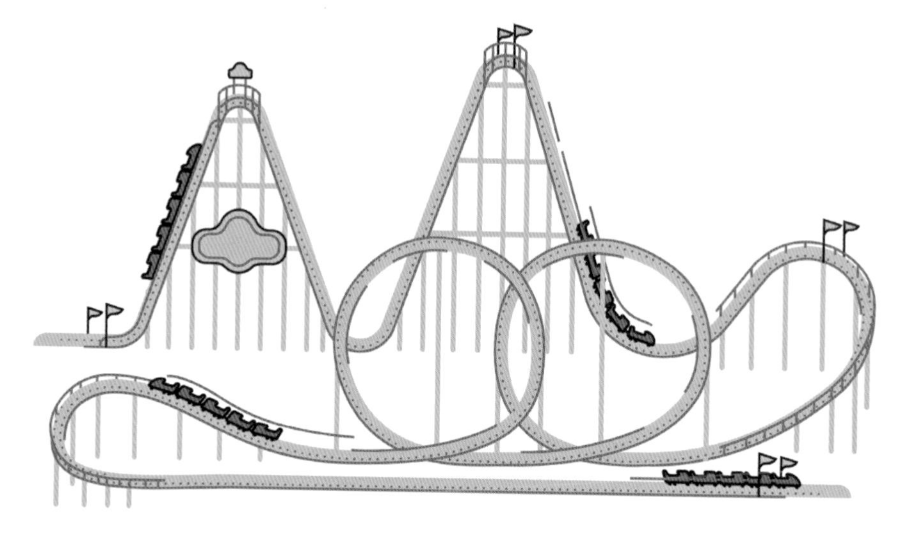

ADVENTURE, AMUSEMENT, AND THEME PARKS

Bless whoever came up with the idea of cramming every adventure possible into a single park. Here are a few that are worth booking a trip for, from fairy-tale wonderlands to worlds of adrenaline-charged adventure.

VISIT THIS EPIC OUTDOOR ADVENTURE PARK IN AUSTRIA

Area 47—Tyrol, Austria

Area 47 in the Oetztal Valley is the largest adventure park in Austria and the ultimate giant playground for thrill seekers and adrenaline junkies. The park is separated into distinct areas where you can have just about every outdoor adventure imaginable: there's an area for rafting and canyoning, a free-ride area for mountain biking, a water area for cannonballing and steep slides, and an epic wakeboarding area. And we haven't even scratched the surface of what's there. The climb area is maybe the most thrilling, with an extremely high ropes course, a mega swing, a flying fox, and bungee jumping from the highest pedestrian bridge in Austria. And if *that's* not exciting enough, swing out over the treetops after a terrifying 148-foot free fall on their monstrous Valley Swing. Is your heart racing yet?

EIGHT ROLLER COASTERS THAT SHOULD BE ON EVERY THRILL SEEKER'S BUCKET LIST

1. **FORMULA ROSSA AT FERRARI WORLD—YAS ISLAND, ABU DHABI, UNITED ARAB EMIRATES**

   Strap in, kids: this is the fastest roller coaster in the world. You'll travel from 0 to 149 miles per hour in just 4.9 seconds and be blasted to a height of 172 feet (52 meters). It's like riding in an F1, but faster and way more intense.

2. **STEEL VENGEANCE AT CEDAR POINT—SANDUSKY, OHIO, UNITED STATES**

   This absolute beast of a ride is a high-speed, hyper-hybrid creation with heart-stopping twists and turns, with a 200-foot (61-meter), 90-degree drop and more airtime than any roller coaster in the world.

3. **THE DEMON AT TIVOLI GARDENS—COPENHAGEN, DENMARK**

   As you hurtle through the loops of the Demon, you'll feel like you're riding on the back of a dragon thanks to this coaster's epic virtual-reality addition.

4. **GRAVITY MAX AT LIHPAO LAND—HOULI DISTRICT, TAICHUNG, TAIWAN, CHINA**

   The world's only "tilt" roller coaster has a piece of track that seemingly detaches midway through, tilts 90 degrees, and then reattaches to drop the car vertically and carry on. Absolutely terrifying.

5. **ALPINE COASTER AT GLACIER 3000—NEAR MONTREUX, SWITZERLAND**

   Zoom across the snow in this unique toboggan roller coaster in the Alps. Your toboggan has a brake so you can control the speed of the drops.

6. **KINGDA KA AT SIX FLAGS GREAT ADVENTURE—JACKSON, NEW JERSEY, UNITED STATES**

   The tallest roller coaster in the world (at 465 feet, or 172 meters!) is also the fastest in North America. Shoot up at 90 degrees and drop back down in an insane 270-degree spiral.

7. **STEEL DRAGON 2000 AT NAGASHIMA SPA LAND—KUWANA, MIE PREFECTURE, JAPAN**

   This giga coaster is one of only seven in the world with a drop of more than 300 feet (91 meters) and holds the record for the longest roller coaster in the world, with a track length of 8,133 feet (2,479 meters).

8. **FALCON'S FLIGHT AT SIX FLAGS QIDDIYA—RIYADH, SAUDI ARABIA**

   Forget the above records; Falcon's Flight is set to open in 2023 and smash all of them, becoming the most extreme roller coaster in the world. Look forward to a height of 525 feet (160 meters) and speeds of 155 miles per hour.

GO ON THE WORLD'S LONGEST CABLE-CAR RIDE

Sun World—Da Nang, Vietnam

Sun World is a huge mountain resort and amusement park complex located up in the hills just west of the coastal city of Da Nang in central Vietnam. With seven amusement parks rolled into one, it's hard to know where to start here. Ride on the world's longest single-rope cable-car system over mountains, rice terraces, and the sea. Take in the amazing views from a 377-foot Ferris wheel. Visit cultural and spiritual attractions galore, get lost in a garden maze, go on thrilling rides, and dance on the streets with food and beer at the Land of Festivals. This place is a lot, and it's amazing—you might decide to never leave.

BRAVE THE JUMPS AT THIS CANYONING WATER PARK

Canyoning Park—Argelès-sur-Mer, France

If you want to experience the thrill of canyoning in a controlled environment, this unique water park in the South of France is a great place to do it. Canyoning Park has re-created a natural canyon environment that's perfect for a day of thrilling water-based adventures, no matter your age or fear tolerance. The family-friendly park has fourteen pools, cliff jumps that range from ten to twenty-three feet, underground river chutes, a zip line, a rope ladder, and more. Rappel through waterfalls, climb to the top of the cliffs, and then waterslide back into the plunge pools. You'll want a decent level of fitness and bravery to participate, but if you prefer to sit back and watch, they do have a café.

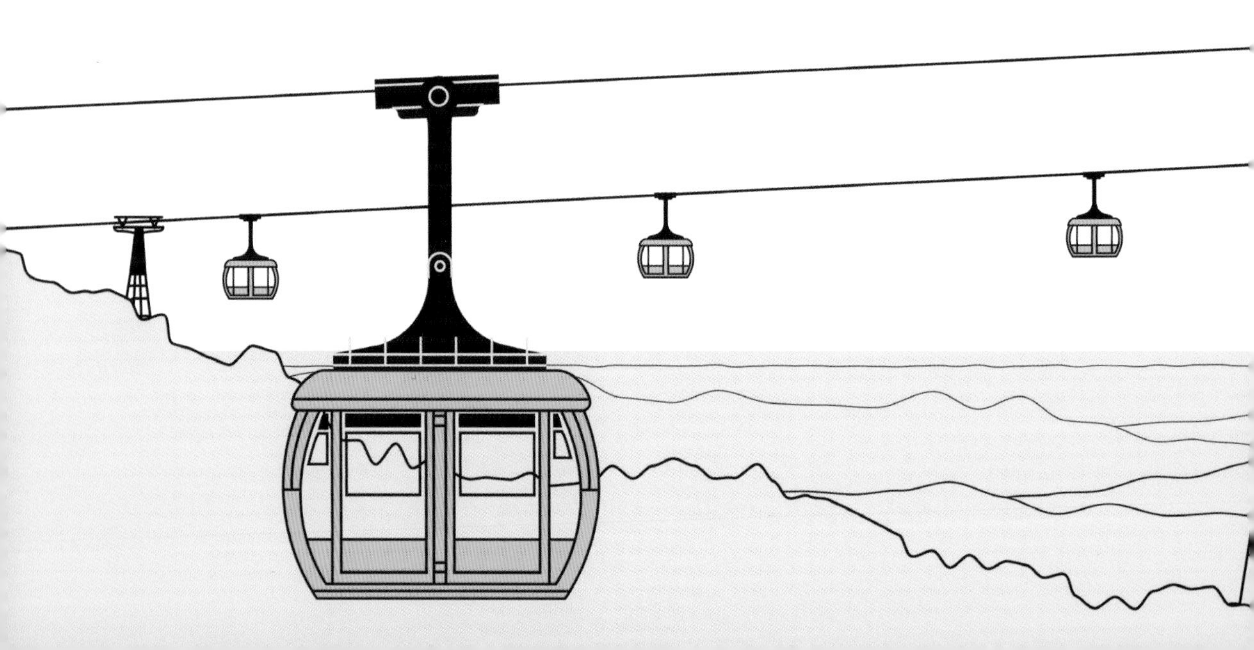

THIS ADVENTURE PARK IN MEXICO OFFERS A FAKE BORDER-CROSSING EXPERIENCE

Parque EcoAlberto—Cajon, Mexico

This eco-adventure park in the central Mexican state of Hidalgo offers kayaking, zip-lining, extreme rappelling, hot springs, and a water park. But maybe their most unusual attraction is a fake border-crossing experience. For three hours at night you'll re-create an illegal border crossing from Mexico into the United States, complete with sirens, dogs, terrifying chases, and the shouts of border-patrol agents. The park is part of the Indigenous HñaHñu community, and the concept came about with the goal of dissuading the young people in the community from trying to make the crossing into the United States themselves. But visitors from all over Mexico and beyond now visit to partake in this anxiety-inducing experience.

LOSE YOURSELF IN AN INTERACTIVE FANTASY RPG EXPERIENCE

Evermore Park—Pleasant Grove, Utah, United States

Lovers of Dungeons & Dragons and all things RPG, this theme park is for you. Evermore is an immersive experience park, dubbed the "world's largest stage." Enter the park and be transported into what feels like a video game in a magical European hamlet, where quest givers assign you tasks and send you on fantastic adventures. It's basically an interactive theater production or live-action role-playing event full of spooky characters, enchanting scenery, fun games and activities, a battle arena, and definitely real dragons. Each season they feature a different themed adventure, so you'll never run out of things to explore.

HAVE A PERFECT DAY OF ADVENTURES (OR CHILLING) ON THIS PRIVATE ISLAND IN THE BAHAMAS

Perfect Day at CocoCay—Berry Islands, Bahamas

CocoCay is a water park, adventure park, beach club, and fantasy island oasis all rolled into one. It's home to Daredevil's Peak, the tallest waterslide in North America—on which you can look forward to a thrilling thirty-five-second watery plunge to the bottom at eighteen miles per hour. Attractions include a giant rainbow helium-balloon ride that takes you 450 feet (137 meters) into the air, a swim-up bar in an oasis lagoon, a beach club with private cabanas, and lazy beaches to chill on with powdery white sands. There's truly something for the whole family here. The island is privately used by Royal Caribbean, and you can book to visit as part of one of their cruises through the Bahamas.

VISIT THE THEME PARK ON PABLO ESCOBAR'S OLD ESTATE

Hacienda Nápoles—Puerto Triunfo, Colombia

Did you know that the luxurious former estate of notorious drug lord Pablo Escobar has been transformed into a family-friendly theme park? After Escobar was killed, the land was reclaimed by the government, which now leases it to Parque Temático Hacienda Nápoles. They have water attractions, a zoo, and giant concrete dinosaurs that Escobar himself installed on the property. And then there are the hippopotamuses. Escobar imported four of them in the 1980s for his private zoo, and they bred into a family of dozens. The park now cares for some of them; others have gone rogue and live in the surrounding area. The whole thing is surreal.

STEP INTO A FAIRY-TALE WORLD OF WONDERS

Efteling Amusement Park—Kaatsheuvel, Netherlands

If you love classic fairy tales, this theme park is for you. Efteling is like the Dutch version of Disneyland; it's a huge fantasy- and folklore-themed park with something for everyone. You can enter an enchanted forest, visit Hansel and Gretel's gingerbread house, and relax while a big talking tree tells you stories. Visit Sleeping Beauty, or float past flowerbeds on a gondola. But it's not all sweetness and magic—they have thrilling adventure rides as well, like a dive coaster that sends you on a free fall into a mine shaft. It all amounts to the perfect mix of fantasy and terror, and it's only an hour's drive from Amsterdam.

EXPERIENCE UTTER MAYHEM AT THIS ANNUAL SCREAM ATTRACTION

Farmageddon—Lancashire, England

Farmer Ted's Adventure Farm offers wholesome day of fun adventures for the entire family. But forget that; that's not why we're here. Each year for the month of October, after dark, the farm is host to a truly terrifying scare attraction called Farmageddon. Four different haunted houses take visitors through a series of horror-filled experiences: think clowns, zombies, and getting locked in a meat locker with an evil spirit who is hungry for flesh. Fun times. There are also fairground rides, bars with bloodcurdling entertainment, and scary-oke (scary karaoke). If you love Halloween season, you won't want to miss this event.

AMAZING BARS

If your idea of thrill seeking involves consuming several adult beverages in a unique setting, you're not alone. To get in the mood, pour a drink and browse this collection of fun-filled watering holes—some are set in incredible locations, others serve up tasty drinks with a side of adventure, but each one has its own refreshing twist.

SIP COCKTAILS BY THE WORLD'S BIGGEST ROOFTOP INFINITY POOL

SkyPark at Marina Bay Sands—Singapore, Singapore

Towering fifty-seven levels aboveground and stretching almost 500 feet (150 meters) across, the rooftop infinity pool at Marina Bay Sands, one of Singapore's most luxurious hotels, is the largest in the world. With unrivaled views of the Singapore skyline, it's fairly easy to laze away the day sipping cocktails poolside. And once the sun starts to set, it'll be even harder to leave. (Thankfully, the pool's open until 10:00 p.m.)

FROLIC IN LONDON'S FIRST BALL-PIT BAR

Ballie Ballerson—London, England

Relive your childhood with added alcohol at this ball-pit cocktail bar in Shoreditch, London. A series of rooms are filled with more than a million balls, which glow like magical, candy-colored orbs thanks to the venue's creative lighting. A general admin ticket will get you into the bar and allotted access to the pits, while the VIP table packages ensure you a private spot with some booze included. As you might expect, the cocktails are fun and nostalgia-inducing, too, with varieties like Hubba Hubba Bubba (bubblegum, lime, and tequila), Skittle Sour (pink gin, limoncello, Skittles), and Caprisunha (watermelon vodka, passionfruit, pineapple). When you get hungry, there's a menu of Neapolitan pizzas sure to fill you up.

HOW TO SAY "CHEERS" IN SIXTEEN LANGUAGES

The act of toasting before a drink is a social tradition that exists around the world. So next time you're at a bar overseas, whip out these phrases and make friends with the locals.

Spanish: "Salud!" *Translation: Health!*

Japanese: "Kanpai!" *Translation: Dry the cup!*

Italian: "Salute!" *Translation: Health!*

German: "Prost!" *Translation: Cheers!*

Chinese (Mandarin): Gānbēi! *Translation: Dry cup!*

Russian: "Za vashe zdorovye!" *Translation: For your health!*

Korean: "Geonbae!" *Translation: Empty glass!*

French: "Santé!" *Translation: Health!*

Arabic (Egypt): "Fe sahatek!" *Translation: Good luck!*

Swedish: "Skål!" *Translation: Cheers!*

Greek: "Yamas!" *Translation: Health!*

Portuguese: "Saúde!" *Translation: Health!*

Afrikaans: "Gesondheid!" *Translation: Health!*

Polish: "Na zdrowie!" *Translation: To health!*

Hebrew: "L'chaim!" *Translation: To life!*

Thai: "Chon gâew!" *Translation: Hit glasses!*

STAY COOL AT THE WORLD'S ONLY ICE BAR ON A BEACH

Ice Barcelona—Barcelona, Spain

You might be familiar with the concept of ice bars, but have you ever heard of one on a beach? At Ice Barcelona, you can get some reprieve from the hot Spanish sun by venturing into the frosty bar, where temperatures hover around 7 degrees Fahrenheit (-14 degrees Celsius). The lounge is fully decked out with ice walls, furniture, and sculptures, and the theme of the decor changes every year. Once you've had your fun—or get too cold—you can head out to the beachside terrace bar to thaw out in the sun.

CALL THE SHOTS AT THIS EPIC BAR IN BUCHAREST

Shoteria—Bucharest, Romania

Shoteria is a super-fun bar in Romania's capital that serves only shots—and they have more than sixty kinds. Drink to science from tiny shot beakers, or try a shot served in a tiny ice goblet. Watch in awe as the bartenders do magic tricks, then throw back another shot because why not? They've got minty shots, fruity shots, creamy shots, coffee shots, even shots arranged in a giant tower. They've got buckets of shots or hollowed-out pineapples full of, you guessed it, shots. Do you get the idea? You'll be drinking a lot of shots here. But don't worry; they make the shots less "strong" so you can keep at it all night. They even serve juicy hangover pickles to help you replenish your electrolytes, too.

GET CRAFTY (AND A LITTLE TIPSY) AT A DIY BAR

DIY Bar—Portland, Oregon

Wielding a hammer while drinking beer? It doesn't get more thrilling than that. Portland's DIY Bar is the ideal place to get boozy and crafty at the same time. On arrival you can choose what project you want to make—options include things like leather koozies, macramé plant hangers, and bracelets. They provide all the tools, materials, and instructions you need, so all you have to do is grab a drink and get started. It's super chill and you can work at your own pace, but if you get stuck, the friendly staff is there to help.

LET FATE CHOOSE YOUR DRINK AT THIS TAROT BAR IN PRAGUE

The Alchemist Bar—Prague, Czech Republic

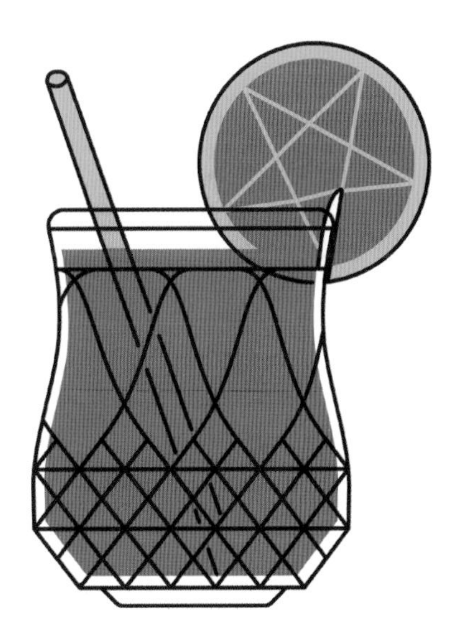

If boozy #magic is what gives you a thrill, visit this mystical bar in Prague that serves classic and contemporary cocktails with an alchemical twist. Choose a card from their Choose Your Fate tarot deck, and not only does this determine what drink you receive, but you'll also be given an instruction or clue that leads you one step closer to solving a mystery and claiming their elusive treasure, which is a prize of about four thousand euros. But if you don't want some divine force deciding your drink, it's cool; you can also just order off the menu.

SIX AMAZING BARS THAT ARE BUILT INTO CLIFFS AND CAVES

1. **SKY LAGOON—KÓPAVOGUR, ICELAND**

   The Sky Lagoon—another geothermal spa in Iceland with incredible ocean views—has a small bar tucked away in a cave at the edge of the lagoon.

2. **CAVE BAR—PETRA, JORDAN**

   Supposedly the oldest bar in the world, the famous Cave Bar is right in front of a two-thousand-year-old Nabatean tomb and surrounded by original carved stonework.

3. **BAT BAR AT LOST CANYON CAVE—RIDGEDALE, MISSOURI, UNITED STATES**

   The Bat Bar is hidden in a cave along the Lost Canyon Cave & Nature golf-cart trail at the Big Cedar Lodge wilderness resort. Pull up in your cart and sip a beverage or two before you continue exploring.

4. **CAVERNA RESTAURANT AT YUNAK EVLERI—GÖREME, TURKEY**

   A restaurant and bar at the incredible Yunak Evleri caves hotel. It's built into the hills and caves of Göreme in Turkey's magical "fairy chimney" Cappadocia region. You can read more about Cappadocia on page 81.

5. **BUZA BAR—DUBROVNIK, CROATIA**

   This iconic bar in Old Town Dubrovnik is built into a cliffside with amazing views and is accessible through a literal hole in the wall.

6. **ALUX RESTAURANT & LOUNGE—PLAYA DEL CARMEN, MEXICO**

   A massive bar and fine-dining restaurant that's built into a ten-thousand-year-old cavern, complete with stalagmites, stalactites, and tasty cocktails.

SOAK UP THE VIEW AT THE ROOFTOP BAR THAT WILL RUIN YOU FOR ALL OTHER ROOFTOP BARS

Sky Bar—Bangkok, Thailand

There are rooftop bars, and then there's Sky Bar. Set 820 feet (250 meters) in the air atop the Lebua Hotel, Sky Bar is the tallest rooftop bar in Bangkok and provides spectacular views of the city and the Chao Phraya River below. The setting might look familiar to movie buffs; the bar featured in *The Hangover II* (if you're feeling adventurous, order the Hangovertini, a cocktail of scotch, vermouth, rosemary syrup, and apple juice that was supposedly created for the film crew). One word of warning: the drink prices are almost as high as the bar itself, so unless you're feeling like a baller, just come for one and enjoy the view—then go back to drinking one-dollar Changs at your Airbnb.

Travel Tip: If you want to get even *higher* (on the views), travel up to the seventy-eighth floor of the nearby King Power Mahanakhon building in Bangkok for their Glass Tray experience. It's as terrifying as it sounds: a glass tray juts out the side of the building from the observation deck, which you can stand on for 360-degree views of the city. There are no drinks here, though.

FLOAT IN A GEOTHERMAL LAGOON WITH A GLASS OF WINE

Blue Lagoon—Grindavík, Iceland

Blue Lagoon is a unique retreat and spa experience in a fishing town about an hour outside of Reykjavík, Iceland, and it's centered around a huge man-made geothermal lagoon. Reserve a day pass for the spa to pass the time in warm, ice-blue volcanic waters with a drink in your hand; swim up to a bar that serves wine, beer, cocktails, and other, nonalcoholic refreshments. You can also get an in-water massage or slather your face in silica mud and algae at the mask bar. It's an invigorating, luxurious experience.

Psst! For more about natural springs and volcanoes, flip to page 51.

This Iconic Jungle Nightclub in Brazil Is Recovering from a Terrible Cyclone: Green Valley is an incredible indoor-outdoor nightclub in the rain forest near the Brazilian resort city Balneário Camboriú, and over the years it has been rated again and again as the best nightclub in the world. But in 2020 (which was already the Worst Year Ever for nightclubs), the beloved club was devastated by a cyclone that almost completely destroyed it. Green Valley plans to rebuild and rise up again better than ever; you can follow them on Instagram for updates: @greenvalleybr.

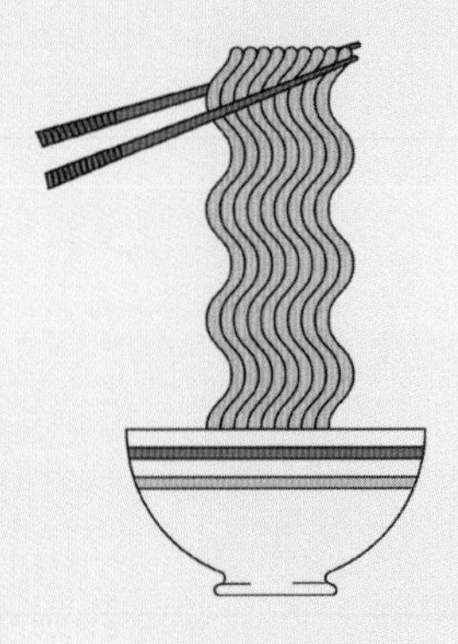

PART FOUR

TASTE TRIPPER

If you're one of those people who plan their travel around their meals, you're in the right place. You already know that food is a way to connect—with places, cultures, people—even if you don't speak the same language.

When you're traveling far from home, it's easy to default to eating foods that feel familiar, comforting even. But, if you can, try to be adventurous and taste as many new things as possible! At worst you might end up with a mouthful that's hard to swallow, but at best you could just discover a new favorite food. (That's worth the risk, if you ask us.)

With so many incredible cuisines around the world, there's no way we could cover them all in the following pages. So instead, we've picked a few out-of-the-ordinary food experiences that are worth jaunting across the globe for. After all, what better reason to travel than for food?

FOOD FESTIVALS

Food and festivals go hand in hand (and not just because of the alliteration). From the world's biggest open-air pizza party to a celebration of canned meat, get ready to nerd out at some rather specific food events that are worthy of celebration.

VISIT THE HOME OF BIBIMBAP

Jeonju Bibimbap Festival—Jeonju, South Korea

Bibimbap, a Korean dish of rice mixed with vegetables and meat, is the star of the show at this festival. It takes place in Jeonju in western South Korea, which is thought to be where the popular dish originated. The town celebrates by assembling a ginormous version of bibimbap—we're talking one big enough to serve more than four hundred people—and mixing it using equally giant wooden paddles. Along with the opportunity to try loads of regional variations of bibimbap, you'll also be met with plenty of live performances, market stalls, and games.

SAMPLE SPAM LIKE YOU'VE NEVER HAD IT BEFORE

Spam Jam—Honolulu, Hawaii

If you're not a fan of Spam, this food festival is here to change your mind. Waikiki Spam Jam is an annual celebration of the polarizing canned meat. Local restaurants take over one of Honolulu's most notable streets to serve up their best Spam dishes to residents and tourists alike: think everything from Spam curries to tacos to musubi, a popular Japanese-influenced Hawaiian dish that draws inspiration from sushi. Along with, well, lots and lots of Spam, you'll also find live entertainment, merch, and Hawaiian crafts. Oh, and entry is free!

Fun Fact: Hawaiians eat approximately seven million cans of Spam each year, making Hawaii the state in the US with the highest Spam consumption per capita.

JOIN IN ONE OF THE BIGGEST PIZZA FESTIVALS ON EARTH

Napoli Pizza Village—Naples, Italy

Every year, crowds of pizza lovers from across the globe flock to Italy to attend Napoli Pizza Village. During the multiday festival, local pizzerias set up stalls along the promenade where guests can learn how to make Neapolitan pizza the proper way (yes, there's an art to it!), take part in competitions, witness "pizza acrobatics" performed with dough, and taste test tons of pizzas, obviously. But the main event is the world pizza championships, where judges spend hours sampling slice after slice of delicious pizza in order to determine the winners in a number of different categories. Tough gig, but someone has to do it.

SEE MASTER SHUCKERS AT THE WORLD'S OLDEST OYSTER FESTIVAL

Galway International Oyster Festival—Galway City, Ireland

When you hear "Irish festival," your mind probably goes straight to St. Patrick's Day. But each year, Galway City in the country's west comes alive with an entirely different kind of celebration: an event dedicated to oysters. Across four days, guests can sample Galway's famed oysters and other local seafood before cheering on contestants at the World Oyster Shucking Championships, a prestigious event where participants are judged by speed, technique, and presentation in the hopes of being named that year's champion. The event has been running since 1954 and takes place annually at the end of September to coincide with the region's oyster harvest.

CELEBRATE EASTER WITH A FIFTEEN-THOUSAND-EGG OMELETTE

Giant Omelette Festival—Bessières, France

Every Easter in the southern French town of Bessières, a huge pan is set up in the town square in preparation for a most *egg-cellent* celebration (sorry, had to!). Hosted by an organization known as the Brotherhood of the Omelette, the event sees thousands of eggs cracked, poured, and mixed up into a massive omelette, which is then handed out to spectators for free. The tradition began in the '70s and has grown in popularity (and omelette size) since then—numerous French-speaking cities across the world also cook up their own giant omelettes.

TRY THE BEST POUTINE IN CANADA

La Poutine Week—Various Locations, Canada

Every year for a week in February, restaurants from across the world gather for the ultimate Canadian cook-off. Poutine typically consists of french fries topped with cheese curds and gravy, but during this festival, restaurants are challenged to put their own spin on the dish and create a unique version (think poutine bibimbap and sweet variations). Festivalgoers can sample all kinds of original poutines and vote for their favorites to win. And the good news is it's hosted in multiple cities across Canada, so you'll have plenty of opportunities to get your poutine fix.

ENJOY GARLIC EVERYTHING

Isle of Wight Garlic Festival—Isle of Wight, England

Garlic enthusiasts, we've found your paradise: this festival on the Isle of Wight in the English Channel is an all-in celebration of your favorite stinky bulb. You can sample unusual treats like garlic beer, black-garlic ice cream, and garlic fudge from the many food stalls, as well as shop for local produce, play games, and see live music and other performances. Conveniently located right near a garlic farm, the event has been running for decades and is usually held every August.

Psst! Need more garlic? Flip ahead to page 184 to find a garlic-themed restaurant.

UNLEASH YOUR SWEET TOOTH AT A CHOC-FILLED FESTIVAL

Salon du Chocolat—Paris, France, and Various Locations

Of course we have to end this section with something sweet: the world's largest chocolate and cocoa event, Salon du Chocolat. It started in Paris in 1994 and has since expanded to a number of cities across the world. Every year, producers and chocolatiers from all over gather for a celebration of all things chocolate. From talks and demos with top chefs to pastry workshops to a chocolate fashion show and more tastings than you can handle, this is *the* event for cocoa lovers. Unsurprisingly, it's also known to get quite busy (who knew there were so many chocolate fiends out there?), so just be prepared to brave the crowds.

EAT AND DRINK AROUND THE WORLD AT EPCOT

Epcot International Food & Wine Festival at Walt Disney World Resort—Orlando, Florida, United States

If you like your food with a side of fun, this festival at Disney's Epcot theme park more than deserves a spot on your bucket list. You can take a taste-bud trip around the world simply by strolling through the park, eating and drinking cuisines from six continents. The exact lineup changes every year, but expect tasty dishes like *empanadas de barbacoa* and guava margaritas from Mexico, hot sake and sushi from Japan, and fish and chips and dark ales from England. It usually runs for a couple of months, so you'll have plenty of time to check it out.

FIVE EXTRAVAGANT FOODS YOU SHOULD TRY AT LEAST ONCE

If you like eating things that are a bit extra, check out these five foods that are nothing short of over-the-top.

1. CAVIAR SCRAMBLED EGGS

Petrossian—Los Angeles, California, United States

Petrossian has been serving fine caviar since their founding in Paris in 1920, and if you have a few thousand dollars to spend, their restaurant and boutique in West Hollywood has a dizzying selection of caviar to choose from. But if you're looking for a more low-key luxury dining experience, their lunch menu is where it's at. Try the creamy scrambled eggs topped with caviar and garnished with fresh herbs.

2. LOBSTER ICE CREAM

Ben & Bill's Chocolate Emporium—Bar Harbor, Maine, United States

So, you love lobster, and ice cream, and always wondered what they'd taste like together? Enter Ben & Bill's, a family-owned shop in Maine serving fine chocolates, candies, and fresh ice cream. Here you can sample their famous Lobster Ice Cream: butter-flavored ice cream with chopped, buttered lobster folded through. Try it in a waffle cone, or, if you really want to commit, grab a half-gallon tub to go.

3. TRUFFLE MAC-AND-CHEESE BURGER

Callahan's—Norwood, New Jersey, United States

Head to Callahan's in New Jersey for their mouthwatering, world-famous truffle mac-and-cheese burger. Is it a bit much? Yes. Will we eat it? Big yes. Smoked gouda, bacon, and truffle mac and cheese is stuffed between two meat patties, breaded, and then deep fried. It's smothered in gooey melted gouda cheese with bacon bits and served on a sourdough roll with lettuce, tomato, and pickles. Follow them on Instagram for their secret fries menu and other seasonal updates (we're patiently waiting for the return of the Flamin' Hot Cheetos–breaded soft-shell-crab sandwich!).

4. WAGYU, FOIE GRAS, AND TRUFFLE CHEESESTEAK

Barclay Prime—Philadelphia, Pennsylvania, United States

We all know Philly has the best cheesesteaks, but this luxury boutique steakhouse in Philadelphia has the fanciest, most expensive cheesesteak at $130 a pop. It's got wagyu rib eye, foie gras, onions, and truffled cheese on a fresh-baked sesame roll. And it's served with a half bottle of champagne, because of course. If you really want to be extravagant, try the lobster cappuccino with porcini and scallop mousseline as a starter.

5. LOBSTER, CAVIAR, AND TRUFFLE PIZZA

Steveston Pizza—Richmond, British Columbia, Canada

This pizza place serves the usual picks: cheese, pepperoni, margarita. But the deeper you get into the menu, the more interesting (and expensive!) things get. There's a section of pizzas inspired by different countries and cuisines and a section inspired by each of the four elements; the "Fire" has pepperoni, capocolli, roasted pimiento peppers, and fresh jalapeños. Then there's the $120 "Hurricane" pizza with shrimp, prawns, eight Canadian lobster tails, and smoked salmon. But if you're really in a mood for extravagance, you've gotta go for the C6: tiger prawns, lobster ratatouille, smoked steelhead trout, and Russian Osetra caviar, all sprinkled with Italian white truffles, all on one pizza, and all for $850.

MEMORABLE EATING EXPERIENCES

Who needs to buy souvenirs when you can take home memories of food? From the perfectly simple to the ones that put the "extra" in "extravagant," you'll be reminiscing about these dining experiences for years to come.

FINE DINE IN A HOT-AIR BALLOON RESTAURANT

Culiair Skydining—Various Locations, Netherlands

Taking to the sky in a hot-air balloon is one thing, but doing so while simultaneously eating a meal worthy of a five-star review? Sounds too good to be true, but it's a real thing, we swear! In the Netherlands, you can float among the clouds in the world's only hot-air balloon restaurant, Culiair. And if that concept isn't quite cool enough on its own, the dishes are literally *cooked inside the balloon*. Food is placed in steel oven baskets before being raised into the balloon via an inventive pulley system. The heat from the balloon cooks the food, which is then transported back down to the kitchen, where it's plated and served to guests. As you'd probably expect, the experience doesn't come cheap, but it's totally *chef's kiss*.

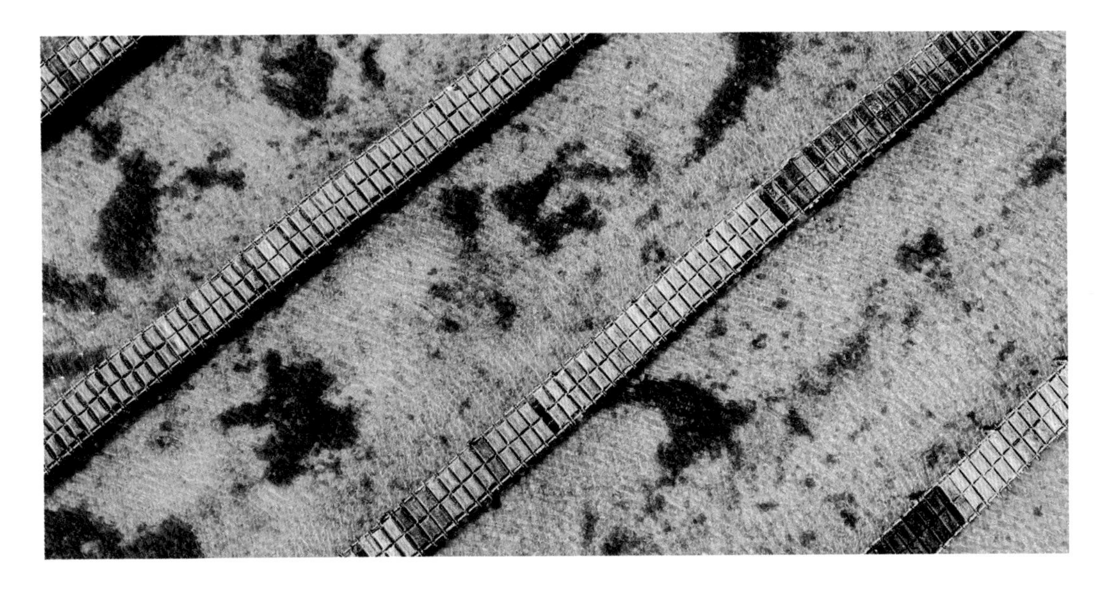

EAT FRESH OYSTERS DIRECTLY FROM THE OCEAN

Oyster Farm Tours—Coffin Bay, Australia

Move over, oceanside seafood restaurants—in Coffin Bay, South Australia, you can sit at a tasting table submerged in the ocean and sample fresher-than-fresh oysters straight from the water. On the tour you can meet the farmer and learn about the history of the area, how oysters grow, and how to shuck them yourself, all while sipping on a glass of local wine. The dress code is strictly waterproof waders (but don't worry, they're included in the ticket cost).

HAVE LUNCH AT A RESTAURANT SET IN A SACRED TREE

Trout Tree—Nanyuki, Kenya

Around three or so hours north of Nairobi, Kenya's capital, there's a pretty special restaurant built around an enormous sacred fig tree, or *mugumo* as it's known by locals. The restaurant specializes in trout (no prizes for guessing where it gets its name), which it sources from the ponds below, so you know it's going to be fresh. There are a couple of alternative menu options if you're not into fish, but if you *are*, you'll be pleased to know they also sell takeaway smoked and fresh trout from the farm so you can enjoy it back at your hotel. Regular guests of the restaurant also include colobus monkeys, hyrax, and plenty of local birds—it is in a tree, after all.

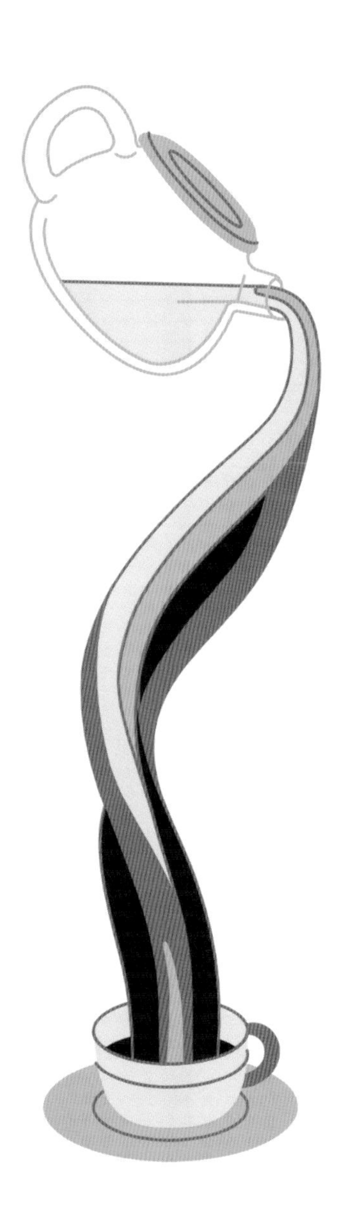

SIP ON COLOR-CHANGING TEA

Chalet da Tia Mercês—The Azores, Portugal

As soon as you arrive in Furnas, a town in the east of São Miguel island in the Azores, the geothermal activity will be obvious: bubbling hot springs, the pong of sulfur, and free-flowing fountain taps that spurt out warm, volcanic springwater. The rustic teahouse Chalet da Tia Mercês makes the most of these natural resources for a truly magical experience. Stop by for a pot of green tea brewed with the mineral-rich water and watch as it reacts with the iron and turns a deep purple color, or try a cup of São George barley “coffee” or perhaps some naturally carbonated lemonade. They also serve snacks made from local Azorean produce. The menu changes seasonally, but you can expect things like bread and sweets baked in geothermal ovens and eggs boiled in nearby hot springs.

EAT WITH THE FISHES AT AN UNDERWATER RESTAURANT

Ithaa Undersea Restaurant—Rangali Island, Maldives

Nothing screams luxury like a meal at an underwater restaurant. Ithaa Undersea Restaurant, part of Conrad Maldives Hotel, is set about sixteen feet (five meters) underwater and has a domed glass roof that offers incredible coral-reef views. They offer fancy set menus for lunch and dinner, which are exxy, but also offer you the chance to spot sharks and stingrays (!) swimming by as you dig into your meal. If you want a taste of the experience without the price tag, opt for the slightly more affordable champagne or brunch package instead. The intimate dining room has space for just fourteen diners, so reservations are a must.

TAKE A CHOCOLATE- AND CHEESE-FILLED TRAIN RIDE

Chocolate Train—Montreux, Switzerland

Ah, chocolate and cheese . . . name a more iconic duo. To get your fix of both at once, you can't go past the Chocolate Train, which takes you on a stunning journey through the Swiss countryside, stopping at delicious cheese and chocolate factories along the way. Passengers board in Montreux, a town on the shore of Lake Geneva, before riding the railway—chocolate croissants and hot chocolates in hand—to the medieval village of Gruyère and ending in Broc, at the Callier chocolate factory, of course. If the samples aren't enough of a reason to jump aboard, the scenery is incredible. Come prepared with an empty stomach (and camera roll).

WALK THROUGH THE INDIAN OCEAN TO A RESTAURANT ON A ROCK

The Rock—Zanzibar, Tanzania

Perched on a rock just off the coast of the island of Zanzibar, this little restaurant offers an unforgettable dining experience. Aptly named the Rock, it specializes in seafood (no surprises there) and uses locally sourced ingredients as much as possible—think seafood from the surrounding waters, spices from the island's spice farms, locally grown veggies, and Tanzanian beef. You can access it on foot during low tide or jump in a boat when the tide is high. Nab a table on the outdoor terrace for the best panoramic views.

ENJOY A FILIPINO FEAST UNDERNEATH A WATERFALL

Villa Escudero—Quezon, Philippines

Around two hours by car from Manila, Villa Escudero is a five-star resort with a waterfall restaurant. Tables are set in the shallows alongside the flowing falls, and the buffet of traditional Filipino dishes is served Kamayan style, a form of communal dining in which food is enjoyed utensil-free (read: with your hands). Once you've had your fill, you can explore the grounds: paddle across the calm river on a bamboo raft, stroll through the gardens, or peruse the museum.

TAKE A GOURMET BUS TOUR OF PARIS

Bustronome—Paris, France

City bus tours can get a bad rap, but before you stop reading, hear us out: this isn't an ordinary hop-on, hop-off bus situation. Bustronome is essentially a high-end restaurant, except instead of sitting in a dining room, guests can enjoy their meal on a fancy double-decker bus while touring the sites of Paris. There's a full glass ceiling on the upper deck so you can soak up all the views, and the classic French food is prepared with seasonal produce. (Bonus: Bustronome also operates in London.)

Travel Tip: If you love cooking as much as you love eating, why not take a cooking class in each new country you visit? It's a great way to learn about different cultures, and you'll be able to reminisce about your travels every time you make one of the dishes back home. Look for classes run by locals for the most authentic experience.

STREET EATS

From mouthwatering morsels to unique snacks that will challenge and expand your culinary horizons, street food has it all. Here are just a few of the tastiest or more intriguing street foods to try around the world.

EAT THE WORLD'S MOST AFFORDABLE MICHELIN MEAL

Hawker Chan—Chinatown, Singapore

Singapore is home to the world's first Michelin-starred street-food stall meal, and although Hawker Chan's tender chicken-and-rice dish lost its Michelin status in 2021, we still think it's worth the hype. The best part is that it all costs less than $3. Chef Chan Hong Meng developed his famous recipe decades ago: tender roasted soya chicken, chopped and served with rice cooked in chicken fat. Hawker Chan has become so popular they now have nineteen (and counting!) locations and franchises throughout Singapore, Asia, and Australia. But the location in Singapore's Chinatown Complex Market & Food Centre is the hawker stall that started it all.

TRY A BOWL OF BIANG BIANG NOODLES IN XI'AN

Barley Market Street—Xi'an, Shaanxi Province, China

Biang Biang noodles are named for the "slap slap" sound they make when cooks prepare them, and though they've started to pop up more and more outside of China, Shaanxi Province is where it all started. These hand-pulled noodles are so long and wide that each bowl needs only three noodles. They're served with braised beef, a splash of fragrant beef stock with plenty of spices (like fennel, cinnamon, Szechuan peppercorn, and star anise), vegetables, fresh herbs, sesame oil, and chili oil. Seriously delicious. Among other places, Biang Biang noodles are prepared fresh in the Barley Market Street in the city of Xi'an. The area is home to a large population of Hui people (a Chinese Muslim ethnoreligious group), so the market specializes in a huge variety of flavorful halal street foods.

Travel Tip: If authentic food is what you're searching for, look for restaurants filled with locals, not just other tourists. And once you find a spot you like, try asking the staff for their own recommendations of other places to eat.

VISIT THIS TASTY BAKERY FOR LEBANESE BREAKFAST PIZZA

Abdel Aziz—Beirut, Lebanon

Maybe one of the most famous street foods in Lebanon, *manousheh* (plural: *manakish*) is a flatbread pizza topped with herbs, meats, or cheeses, or a combination. If you're in Beirut, the bakery Abdel Aziz in the Hamra neighborhood is the spot to go for tasty *manousheh* fresh out of the oven. Try it topped with *za'atar* (an aromatic mix of herbs, spices, and sesame seeds), *kishk* (a powdery delicacy of fermented yogurt with bulgur wheat), and cheese, or order *lahm bi ajin* if you'd like ground beef or lamb cooked with tomatoes, onions, spices, and fresh herbs topped on tasty flatbread. You can find *manakish* just about anywhere; just follow their fragrant scent to the nearest good bakery.

STOP BY THIS ICONIC TORTILLA STAND IN SAN DIEGO

Café Coyote—San Diego, California, United States

Out front of Coyote Café y Cantina in the historic Old Town, San Diego, Elida, "the tortilla lady," has a stand where she makes fresh tortillas for the restaurant and for passersby to snack on. She has made more than six million tortillas and counting! They're so incredibly soft and fresh that you can eat them with a bit of salsa or butter only. If you're looking for a bigger meal after your tortilla snack, get a table in the colorful restaurant and sample Café Coyote's delicious Mexican food and epic selection of margaritas. And when you're thoroughly full and satisfied, stop by Elida's stand again for a fresh strawberry- or chocolate-flavored tortilla to go.

SAMPLE FAMOUS MOROCCAN SNAIL SOUP

Jemaa el-Fnaa Square—Marrakech, Morocco

Jemaa el-Fnaa is an important square in the medina quarter of Marrakech. Here you'll find vendors selling *babouche*, or snail soup, a common street food in Morocco. The snails come in a bowl of warm broth flavored with around fifteen herbs and spices; you pick the snails out with a toothpick before you drink the tasty soup. If you can get past the snail's texture, they have an earthy, mushroomy taste, and the soup is fragrant and spicy. When you're done, you won't have to look far to find some delicious nougat and sweet tea with fresh mint to cleanse your palette.

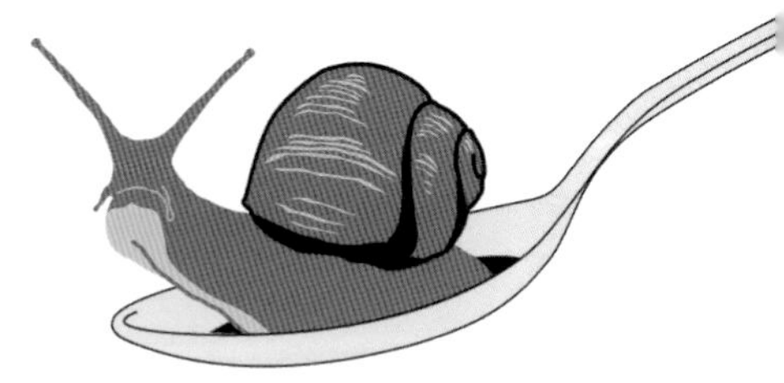

FIVE DELICIOUS STREET FOODS TO TRY IN TAIPEI, TAIWAN

A trip to the capital of Taiwan (China) isn't complete without a stroll along one of the city's snack streets, and Raohe Night Market is an excellent place to start. Here are five foods to try.

1. **Black-Pepper Bun:** A soft bun stuffed with juicy pork, green onion, and black pepper and cooked along the edges of a charcoal-fired tandoor oven. A spicy, peppery night market staple.

2. **Stinky Tofu:** Taiwan's answer to blue cheese. A pungent, fermented tofu, deep fried and topped with spicy, garlicky sauces and fermented cabbage. It's salty, it's acidic, it's smelly—you've got to try it.

3. **Grilled Street Corn:** It's really about the sauce. Fresh corn cobs are smothered in savory, smoky, garlicky *shacha* sauce and then grilled to perfection.

4. **Ice Cream Burrito:** Scoops of pineapple, taro, and peanut ice cream, peanut shavings, and optional cilantro are all rolled into a thin flour crepe, burrito style.

5. **Boba Brûlée:** Fried toast stuffed with crème brûlée and topped with black boba pearls and a Pirouline. Creamy, chewy deliciousness!

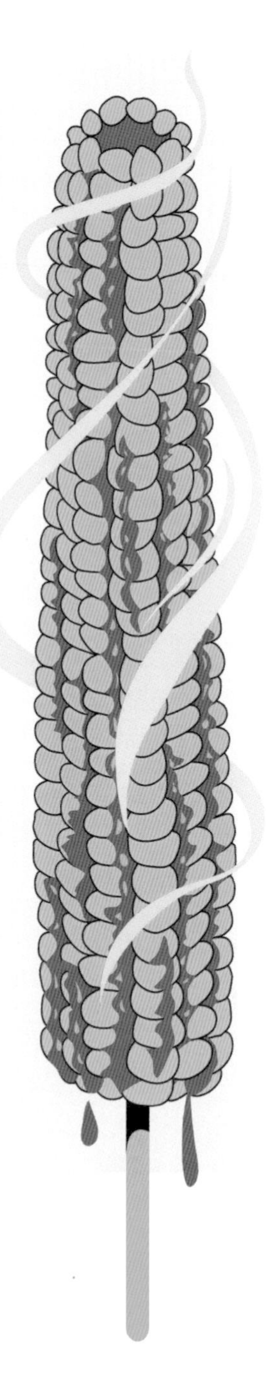

BITE INTO MUMBAI'S BIGGEST SANDWICH

Bipin Big Sandwich—Borivali West, Mumbai, India

Lovers of the humble sandwich: you have to try the Bahubali four-layer sandwich, found in Borivali West in Mumbai. What's in it? A whole lot of veggies (and some fruit for good measure) layered between several slices of crustless white bread. Expect baby corn, pineapple, black olives, jalapeños, tomatoes, and shredded beets, cabbage, mango, and cheese. Szechuan sauce and mayo also make an appearance. There is *so* much happening in this sandwich, but the combo of sweet, spicy, and crunchy between lawyers of soft white bread makes it all work surprisingly well.

SAMPLE LONDON'S BEST STREET FOOD

Street Food Union—London, England

Street Food Union operates the Rupert Street Market in Soho—and it's one of the best places to stop for lunch in London and sample the best street food talent in town. Traders rotate, but you can expect to find mouthwatering eats like fresh vegan and vegetarian falafel wraps, cauliflower shawarma, panko-fried buttermilk chicken, Filipino pork adobo, or Uzbek slow-cooked lamb with spices.

Must-Try British Street Food: The Yorkshire Burrito, which you can find at London's Camden Market, is soft braised beef shin enveloped in a Yorkshire pudding wrap with gravy and cheese. There's also a cauliflower and cheese burrito. It's British comfort food, reimagined.

BUGS, BUGS, AND MORE BUGS

According to the United Nations, we should all be eating more insects if we want to combat global food insecurity and boost nutrition. So with that in mind, next time you see an insect in street-food form, stop and give it a try. Maybe it's about time we stop giving bugs a bad rap.

THE WHO'S WHO OF INSECT STREET FOOD

Apin: Deep-fried tarantulas in Cambodia. Under the Khmer Rouge, people learned to cook these giant spiders to survive. These days, they're a delicacy.

Beondegi: A Korean street food made with silkworm pupae. The pupae are steamed or boiled, seasoned, and served in paper cups with a toothpick.

Chapulines: Mexican grasshoppers, common in the state of Oaxaca. They're fried or roasted and topped with lime and chili or sprinkled dry over other foods for a delicious, crunchy texture.

Escamoles: Fried ant larvae, also known as "desert caviar." The large larvae are fried in butter with spices and are served straight up or in tacos. They're a bit fancy, so you probably won't find them at a typical taco stand, but if you do see them on a menu somewhere, they're definitely worth tasting.

Hachinoko: A dish made of wasp or bee larvae and a delicacy in Japan. Try it at the Hebo Matsuri Wasp Larvae Festival, held every year in the town of Ena, Gifu Prefecture.

Hormigas Culonas: Big-bottomed ants, a delicacy in Santander, Colombia. The best ones are roasted in salt, served up in tiny bags, and eaten like peanuts.

Maeng Da: A popular snack in Thailand. Giant water beetles are fried in oil with chili pepper.

Masonja: Mopane worms, a delicacy in Zimbabwe and other parts of southern Africa. Fried and eaten as a crispy snack.

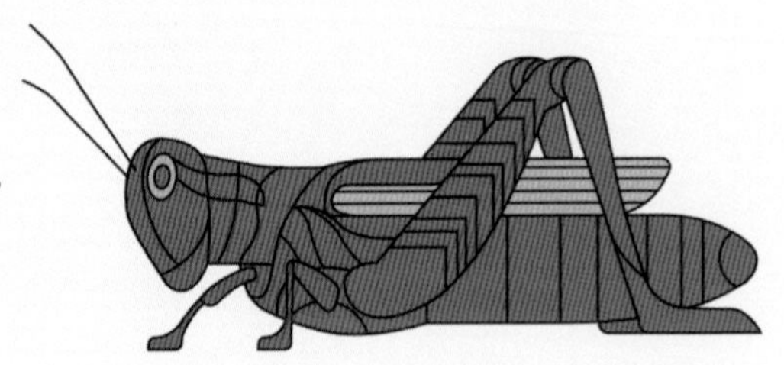

COOK AND EAT BUGS IN CAMBODIA

Fear Factor Challenge—Siem Reap, Cambodia

If bugs make you squeamish, now that you know they're a potentially world-saving delicacy, you're no doubt ready to overcome your fears. In Cambodia, bug street food is *everywhere*—but one place even teaches you to cook them yourself. The Fear Factor Challenge in Siem Reap is a unique cooking experience in which a chef shows you how to prepare tarantulas, scorpions, or crickets, all marinated in delicious spices and fried to perfection—you'll be over your fears in no time. You can book the experience via backstreetacademy.com.

DISCOVER THE ORIGINAL INSTA-WORTHY STREET SWEET

Ci Qi Kou Market—Chongqing, China

Sugar painting is a folk-art practice from Sichuan Province in southwestern China that dates back to the seventeenth century, and in Ci Qi Kou market in Chongqing you can still find street artists making these sweet creations today. Liquid sugar is drizzled onto a hard surface to create intricate designs in the shape of fish, birds, dragons, or butterflies. Spin a wheel to see what shape you'll get, then watch as the artist gets to work, adding a bamboo stick so you can hold your two-dimensional lollipop. They then use a palette knife to lift it up, and voilà—you have a fancy sugar painting to nibble on as you wander the market.

GRAB SOME *ALCAPURRIAS* FOR A BEACHSIDE SNACK

La Alcapurria Quemá—San Juan, Puerto Rico

Alcapurrias—stuffed fritters—are a popular Caribbean snack that are especially good in Puerto Rico. A dough made from green banana and *yautía* (taro root) is filled with savory ground beef and fried to perfection. You'll find *alcapurrias* at any beachside kiosk selling *cuchifritos* (fried goods), and they really are best eaten at the beach, washed down with an icy cold Medalla Light beer. We also recommend ordering them with some smashed sweet plantains at the La Alcapurria Quemá at least once during your trip. Beyond *alcapurrias*, this casual eatery serves up all kinds of flavorful and delicious Puerto Rican dishes, and you'll want to order one of everything on the menu.

DIP INTO PIG'S BRAIN GRAVY

Azul Cebu—Cebu City, Philippines

Tuslob-Buwa (which means "dip in bubbles") is a street food that supposedly originated in the Barangay Suba village in Cebu City, Philippines. This Cebuano specialty combines pig's brain and pork or chicken liver with onions, shrimp paste, and lard, all sautéed in a giant wok. The restaurant Azul Cebu has brought the street food off the street and into an alfresco dining setting, where you actually cook the gravy yourself using a portable stove on the table. To eat, season with soy sauce and chili flakes and serve with fresh *pusô* (hanging rice; it's steamed rice wrapped in coconut leaves) to dip in the flavorful gravy.

Umm . . . Pig's Brain? If you eat pork, don't be intimidated by the idea of pig's brain. In a lot of cultures, an animal's brain is considered a delicacy. Just close your eyes and enjoy the deliciousness, knowing that no part of the animal has gone to waste.

COME FOR A CAR REPAIR, STAY FOR THE *AL PASTOR*

El Vilsito—Mexico City, Mexico

Ask a local where to find the best *al pastor* tacos in Mexico City, and there's a good chance they'll send you to El Vilsito. But don't show up before 8:00 p.m. (unless you need your rental car fixed), because this iconic taco spot is actually a mechanic shop by day. After dark is when it comes to life with people spilling onto the street, filling up on plates piled with perfectly marinated spit-grilled pork folded into corn tortillas with pineapple, cilantro, and diced onions. If you don't eat pork, don't worry; they have other meat options on the menu, too. Stop by the giant *molcajete* bowls filled with fresh green or red salsa, then find a good standing position and enjoy!

Looking for Vegetarian or Vegan Tacos? We recommend visiting Por Siempre Vegana Taquería in Mexico City instead. Flip to page 175 for more info.

NINE GIANT FOODS TO CONQUER IN THIS LIFETIME

Is bigger really better? You be the judge. These monster-size dishes in the United States are perfect for sharing, so grab your hungriest travel buds and prepare to feast.

1. THE OMG BURGER

The Catch—Anaheim, California

Ooft, okay, are you ready for this one? Let's break it down: the OMG burger contains five pounds of ground chuck steak, sixteen (!!) slices of cheese, half an iceberg lettuce, tomatoes, red onion, aioli, and pickles, sandwiched in a fourteen-inch brioche bun. And a whopping side of fries, of course. It's enough to serve eight to ten folks.

2. GIANT CUBAN SANDWICH

Sarussi Subs—Miami, Florida

The Cuban sandwich is a staple in South Florida. Order the Man vs. Food (the original, doubled) at Sarussi Subs, and you'll be served an unforgettable version on a sixteen-inch Cuban sub, loaded with the usual suspects—ham, baked pork, cheese—then topped with pickles and a secret sauce. It weighs around five pounds, so prepare your stomach accordingly.

3. GIANT SOUP DUMPLINGS

Long Xing Ji—San Gabriel, California

The only thing better than xiao long bao, a.k.a. soup dumplings, is giant xiao long bao. This oversize dumpling is filled with a mix of pork and crab and takes up an entire steamer basket to itself. It's served with a boba straw so you can slurp up all the soupy goodness before tucking into the meat.

4. GIANT CANNOLI

Gelso & Grand—New York, New York

Forget saving room for dessert; you're going to want to start off your meal with this extra-large cannoli. The menu changes seasonally, but you can expect options like a wintery hot cocoa cannoli dressed in chocolate, marshmallows, and gold dust, or a unicorn-themed one filled with lavender-berry buttercream and strawberry puree. The fun part? Before eating, you get to smash it up with a mini wooden mallet.

5. A SUPER BIG BOWL OF PHO

Dong Thap Noodles—Seattle, Washington

This version of pho, the classic Vietnamese soup, may be big, but it doesn't compromise on quality. One "Super Bowl" contains about three pounds of meat (a combination of pork brisket, steak, and meatballs), handmade rice noodles, and around six pints (three liters) of broth, made from beef bones and simmered for more than fourteen hours.

6. GIANT SICILIAN PIZZA

Big Mama's & Papa's Pizzeria—Los Angeles, California

This pizza is so big, it doesn't even fit inside the oven; the restaurant had to make an extension so they could cook it with the oven door still open! Measuring 54" x 54", it comes topped with homemade tomato sauce and mozzarella cheese (additional toppings cost extra) and gets sliced up into two hundred squares that can serve sixty to seventy people—or, you know, a few really hungry ones. Did we mention you can get it delivered?

7. A MOUNTAIN OF APPLE PIE

Blue Owl Bakery—Kimmswick, Missouri

How does one make a pie mountain, you ask? First, eighteen apples are peeled, cored, and thinly sliced, then coated in cinnamon sugar. Next, the apple slices are stacked by hand in a bowl, before being turned out into a pie crust and wrapped up with dough, baked, and, finally, topped with a caramel pecan sauce. The result: a beautiful creation known as the Levee High Pie, weighing in at ten pounds and measuring eight inches tall.

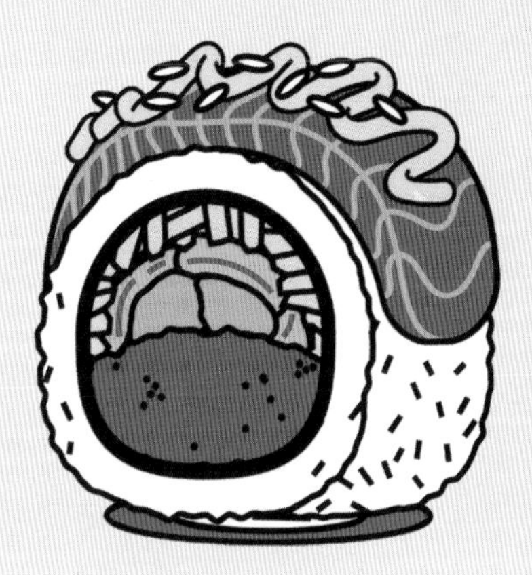

8. SIX-POUND MONSTER SUSHI

Deli Sushi & Desserts—San Diego, California

Any sushi roll at this restaurant can be upsized into a "Monster Roll," which is the equivalent of eight sushi rolls in one! There is a challenge to try to finish eating an entire roll on your own in just fifteen minutes, but we reckon it tastes so good it's worth savoring—and sharing. Either way, you won't leave hungry.

9. GIANT KATSU CURRY

Kenka—New York, New York

Japanese food fans, buckle up. This restaurant in New York's East Village serves up a jumbo curry dish containing three pounds of curry served over three and a half pounds of rice. If that doesn't sound like enough, you can top it with pork katsu (breaded and fried pork cutlet). You know, as a garnish.

Travel Tip: If you have a food allergy and you're traveling to a country where you don't speak the language, write down "I'm allergic to [insert food here]" and find someone (a local or Google) who can translate it for you into the language of the country you're visiting. Then, anytime you go to a restaurant, you can just whip out your note and show it to the server.

DECADENT DESSERTS

Dessertarian travelers know that every good trip needs plenty of sugar stops. And these decadent creations will have you wanting to plan an entire vacation around where to eat something sweet.

TRY THESE HEAVENLY PERSIAN ICE CREAM FLAVORS

Mashti Malone's—Los Angeles, California, United States

Mashti Malone's is an institution in LA; it's a Persian ice cream shop that has been serving those in the know with fragrant, creamy treats since 1980. They offer all the usual flavors and then some, but if you can bypass the "basic" (sorry, Rocky Road and Butter Pecan), you'll find aromatic flavors like French Lavender, Saffron Rosewater, Persian Cucumber, or Orange Blossom and Pistachio. Try a Malone special, a crispy honey and rosewater pastry with a big scoop of ice cream on top, or go for a classic cup with three flavors—trust us, you'll want to try everything. Fortunately, they also sell pints to go.

STUFF YOUR FACE WITH FRESH *STROOPWAFEL*

Rudi's Original Stroopwafels—Amsterdam, Netherlands

If you're visiting Amsterdam for the first time, don't leave without trying freshly baked *stroopwafel*. Two round, deliciously spiced wafers are baked and then pressed together with sweet, sticky *stroop* (a warm caramel-ish syrup) in between. You can even have them dipped in Belgian chocolate for extra decadence. The original and the best are at Rudi's Original Stroopwafels at the Albert Cuyp Market in Amsterdam, and their family recipe has been passed down for generations. Follow the delicious aroma to find their food truck around the center of the market.

Did Someone Say . . . Stroopwafel Milkshake? Bulls and Dogs is a hot-dog bar in Amsterdam that also serves up epic "freakalicious" milkshakes, including a decadent *stroopwafel* milkshake. The Dutch Cookie Wookie has a classic spiced wafer stuffed with caramel sauce blended into a thick shake, with more *stroopwafel* piled on top of whipped cream and drizzled with more caramel. *Mmm.*

DISCOVER COLORFUL MEXICAN *GELATINAS*

GeLATINX—Los Angeles, California, United States

If you grew up in a Mexican family, you probably have memories of *gelatina*—jelly creations that feature layers of fresh fruit, nuts, cake, or other confections suspended in colorful, architectural molds. You'll often see them in shop windows or sold by street vendors in Mexico, but GeLATINX is making *gelatina* to order in the Boyle Heights neighborhood of Los Angeles, many with flavors inspired by nostalgic childhood treats—think a Sponch *gelatina* with delicious marshmallow spongy cookies suspended inside or a Panqué de Elote and Mini Mantecadas mashup sprinkled with crumbled muffin and sweet corn. The Mango Chamoy *gelatina* even comes with tamarind candy and a tiny bottle of Tajín on top. Follow GeLATINX on Instagram to keep up with the latest creations and send a DM to place your order: @gelatinx_la.

JIGGLE THIS PERFECT, PUFFY CHEESECAKE

Uncle Rikuro's—Osaka, Japan

If you've never looked at a cheesecake and thought, "Wow, I want to smack it," then you've been missing out, and Uncle Rikuro's in Osaka makes the fluffiest, jiggliest freshly baked cheesecake you'll ever taste. For those who haven't tried it yet, jiggly cheesecake gives you the creamy flavor of cheesecake with a light, fluffy body that's so much fun to hold, because (you guessed it!) it bounces and wobbles around at the slightest touch. Baked fresh in-house with a layer of raisins on the bottom, the cheesecake is simple and delicious and can be enjoyed to go or in the café.

BITE INTO A COTTON-CANDY ICE CREAM BURRITO

Creamberry—Las Vegas, Nevada, United States

You heard us: this ice cream shop in Las Vegas is famous for their cotton-candy burrito. That's several scoops of ice cream and dry toppings all rolled up inside a rainbow of multiflavored cotton candy. Hello, sweet sugar high! If that sounds like too much, Creamberry also serves puff waffles, rolled ice cream, fancy toasts, and a variety of other sweet treats. On the other hand, if a cotton-candy ice cream burrito just isn't extreme enough for you, you can order a ten-scoop ice cream "bouquet." You deserve this!

JUST THREE EPIC SUNDAES

1. **COCO GELATO—CARDIFF, WALES**

   If Roman emperors ate sundaes, they'd look like the Roman Empire special at Coco Gelato. Four hot waffles are topped with eighteen scoops of gelato, fresh fruit, chocolate bars, fudge brownie chunks, and a selection of syrups and toppings.

2. **MO & MOSHI—BANGKOK, THAILAND**

   Gather your closest to split a twenty-two-scoop sundae at Mo & Moshi. Their Strawberry Supreme is loaded with strawberry scoops, vanilla soft serve, mini waffles, an entire layer of cake, fresh strawberries, and strawberry Pocky sticks. Yum!

3. **MARGIE'S CANDIES—CHICAGO, ILLINOIS, UNITED STATES**

   If twenty-two scoops just won't cut it for you, Margie's Candies in Chicago supposedly serves *the* World's Largest Sundae with a whopping twenty-five scoops. Margie's is an institution—they've been operating for more than a hundred years now, and they're not messing around.

DESSERTS IN LONDON ARE REALLY GOOD

London has an incredible food scene, and one thing the city does well is decadent desserts. Here are three dessert experiences you shouldn't miss.

TREAT YOURSELF TO LUXURY ÉCLAIRS

MAÎTRE CHOUX

If you're looking for a fancy treat in London, you can't get much more decadent than the éclairs at Maître Choux. *Declaired* (pun absolutely intended) as the first choux pastry specialist in the world, Maître Choux is a modern take on a traditional French patisserie founded by Michelin-starred chef Joakim Prat. The patisserie (now with five locations) is best known for its unique takes on the classic éclair, with unique fillings like salted butter and caramel or Persian pistachio mousseline cream. The colorful creations are topped with candied nuts, fruit, flower petals, or even gold powder, and they almost look too pretty to eat. But you'll want to eat multiple.

GET YOUR FINGERS ON THESE DELECTABLE FINGER DONUTS

LONGBOYS

Longboys specializes in handmade finger donuts: delicious long-shaped donuts with fresh fillings. This shop challenges the idea of donuts being calorie laden and overly sweet—their unique creations are lighter and use less sugar throughout the process (but, honestly, you'd never even know). The result is a flavorful, decadent treat that you don't have to feel even the slightest bit guilty about. You can sample quirky and unique flavors like Lemon Meringue Pie, Vegan Peanut Butter Jelly, or Raspberry Rose Lychee, which comes decorated with fresh lychee and raspberry pieces and dried rose petals. Or try their best seller, the Triple Chocolate Brownie: milk chocolate *cremeux*, cocoa crumble, and homemade brownie chunks sprinkled on top.

SIP ON LONDON'S MOST DECADENT HOT CHOCOLATE

DARK SUGARS

You haven't tried chocolate until you've tried it at Dark Sugars, a chocolate shop on Brick Lane that brings the culture of West African cocoa production to London's East End. You'll want to stop here for the handmade chocolate truffles, chocolate pearls, and other rich chocolate goodies, like pop rocks concealed in dark-chocolate shells. They also have a lot of vegan options; try the dried mango dipped in dark chocolate. Don't leave without indulging in one of Dark Sugars' iconic hot chocolates: rich cocoa in flavors like Moroccan chili, cinnamon, or white chocolate, topped with frothy milk and garnished with epic shards of fresh chocolate. It's like chocolate fireworks in a cup.

BITE INTO TWENTY-FOUR DREAMY LAYERS OF CHOCOLATE CAKE

Maison Pickle—New York, New York, United States

Maison Pickle is a contemporary dining spot on the Upper West Side in NYC that specializes in cocktails and French dips—but if you ask us, dessert is the star of the show. They're known for their epic twenty-four-layer chocolate cake: that's twelve layers of decadent cake and twelve layers of smooth chocolate filling for you to enjoy in one big bite. They also have a colorful twenty-four-layer confetti cake—a perfect birthday treat that you should totally order even if it isn't your birthday—and a rich Black and White Oreo Ice Box Pie that will send you into a dessert trance.

TRY SWEET TAPAS AT THIS DESSERT RESTAURANT IN SPAIN

espaiESSENCE—Barcelona, Spain

Espaisucre is a dessert school in Barcelona and claims to have opened the first dessert-based restaurant in the world in 2000. They offer a unique gastronomical experience called Essence, which is like dessert school but for your mouth. Book a ticket and enjoy a surprise degustation menu (which changes seasonally) of three savory tapas, three sweet tapas, and five other desserts. You'll experience unique and surprising flavor combinations and will learn about each of the ideas underlying these creations as you sample them. There are only eight places available for the experience each Saturday, so be sure to book ahead.

WATCH THIS LAVA CAKE OOZE MATCHA DELICIOUSNESS

U Dessert Story—San Francisco, California, United States

We couldn't finish this dessert story without mentioning U Dessert Story in San Francisco: a dessert lover's dream offering Korean-, Japanese-, and Thai-inspired creations made using fresh, organic ingredients. The Mango Sticky Rice Bingsoo is a highlight, with Korean shaved ice, homemade crumble, sticky rice, mango puree, and condensed milk to drizzle all over. Or order a Matcha Lava Cake, break open the soft chocolate exterior, and get your spoon (and camera) ready to capture the delicious green lava that oozes out. You can even spike it with rum shots to make it a boozy treat! They also serve extremely photogenic crepe cakes in matcha or Thai tea flavors and a huge variety of other desserts and milk teas: try the lavender matcha on ice.

BOOZE-INFUSED ACTIVITIES

One of the best things about vacation? It's perfectly acceptable to drink in the a.m. Whether you're a wine snob, a beer connoisseur, or a general appreciator of all things containing alcohol, you can quench your thirst with these boozy experiences. (Please enjoy responsibly.)

TRAVEL ON A TEQUILA TRAIN TO . . . TEQUILA

Jose Cuervo Express—Tequila, Mexico

If this agave-based spirit is your drink of choice, you're going to love this: tequila brand Jose Cuervo has a train that travels between Guadalajara and Tequila, the town where tequila was invented. There are a couple of different train carriages; options range from the Express Wagon (the most basic but also the cheapest and only family-friendly car) to the Elite Wagon (hello, full bar, wood panels, and floor-to-ceiling windows). On board, you'll get to admire the agave fields and sample different tequilas in an expert-run tasting session. The train trip itself takes only about two hours, but the experience also includes a tour of the Cuervo production facilities in Tequila, a Mexican show, and an agave-harvesting demo.

DRINK YOUR WAY ALONG MARGARITA MILE

Margarita Mile—Dallas, Texas, United States

That's right, you can go on a self-paced margarita bar crawl in the heart of Dallas. Despite how it sounds, Margarita Mile isn't a single stretch of road, but rather a cluster of local bars and restaurants across the city. Dallas may claim to be the home of the frozen margarita, but that's not all you'll find; each participating venue has a unique version of the tequila-based drink for your sipping pleasure. From avocado or prickly pear margaritas to frozen ones made with liquid nitrogen, you'll be a marg connoisseur before you can say "Bottoms up!"

SLEEP OVERNIGHT AT A BREWERY

The DogHouse—Columbus, Ohio, United States

Opened by Scottish beer company BrewDog in 2018, the DogHouse claims to be the first beer hotel in the world. It has everything you'd expect: an on-site bar (obviously), views into the brewery, and beer-y aromas from the fermenting foeders. But it's the creative touches—in-shower beer fridges and personal beer taps in every room—that'll blow your beer-loving mind. And the best part? Whenever you're ready to ~~*pass out*~~ sleep, your bed's only a short stumble away. There are plans to open a second DogHouse hotel in Manchester, England, too.

DRINK COCKTAILS BASED ON YOUR MOOD

Bar Bisou—Paris, France

Ever open a menu and see so many options that your eyes completely glaze over? You won't have that issue at this creative little Parisian cocktail bar—menus don't exist here. Instead, the bartender will ask you questions about your tastes, desires, or even your mood and customize a cocktail that's perfect for you at that very moment. Sustainability is also a focus: they take a seasonal approach, sourcing fresh French fruits and vegetables, and minimize waste by using as much of the produce as possible (think edible garnishes made from dehydrated leftover fruits). The space, with its long marble bar, greenery, and pink flamingo–covered bathrooms, is as aesthetically pleasing as the cocktails.

GO FORAGING TO MAKE YOUR OWN HERBAL LIQUEUR

The Drink Workshop—Ibiza, Spain

The famous party island of Ibiza is synonymous with booze, but this alcohol experience might not be what you'd expect. This workshop teaches people how to make *hierbas*—which means "herbs" in Spanish—a centuries-old Balearic liqueur that's often served as a digestif. You'll go foraging for produce like thyme, rosemary, mint, and juniper berries, which you'll then turn into your very own custom blend to take home. The only bad news is you'll need to wait about eight weeks until it's ready to drink, so you'd best sample some from the local bars in the meantime. It'd be rude not to.

SIP ON SAKE AT THE FIRST-EVER SAKE BREWERY IN JAPAN

Sudo Honke—Kasama City, Japan

Sudo Honke has been brewing pure rice sake for more than 850 years, making it the oldest sake brewery in Japan. Their sake is made using high-quality rice (within five months of being harvested) and a traditional method that's been passed down through generations; it's all unfiltered and has no carbonation, producing a well-balanced drop. Book a tour to learn about the brewing process and the history of sake and to sample a couple of different varieties.

CRAFT YOUR OWN GIN AT THIS DISTILLERY HOTEL

The Distillery—London, England

Calling all discerning drinkers: this working distillery in the middle of London's Notting Hill needs to be added to your "places to drink at" list, stat. Enroll in the Ginstitute class to discover the history of gin, sip on cocktails, and create your very own blend to take home. Gin not up your alley? You're in luck—they also offer other spirit-focused experiences, such as the Whiskey Thing and Agave Sessions, as well as special themed nights. And if you really want to commit, book one of the three hotel rooms upstairs, use your newfound knowledge to whip up another G&T at your mini in-room bar, and call it a night.

STOP WHAT YOU'RE DOING: IT'S WINE TIME

If you love wine more than you probably should, you've likely taken part in your share of vineyard tours and tasting sessions. When you want to try something a little different, check out these vino experiences with a twist.

EXPLORE MENDOZA'S WINE REGION ON TWO WHEELS

MENDOZA, ARGENTINA

Sure, going on a bike ride is fun—but it's even better when there's wine at the end. Mendoza sits just near the Chilean border in the heart of Argentina's wine region and is famous for producing some of the finest malbec in the world. The best way to see the area is on two wheels: join a bike tour or hire one (there are plenty of bike rental shops around) and cycle at your own pace from winery to winery. Just try not to drink *too* much, so you can keep your wits about you on the road. If you do have a touch too much, no judgment—that's what taxis are for.

SLEEP IN A GIANT WINE BARREL

QUINTA DA PACHECA—DOURO, PORTUGAL

Located in the Douro Valley, one of the world's oldest wine regions, this historic winery offers guests the chance to sleep in an oversize wine barrel in the middle of the vineyard. It's comfier than it sounds—each barrel is decked out like a luxe hotel suite, complete with a bathroom and private terrace out front. There's plenty to do on the property, too, like wine tastings, tours, spa treatments, and harvest experiences.

ROAD TRIP ONE OF THE LONGEST WINE ROUTES IN THE WORLD

ROUTE 62—SOUTH AFRICA

You'll want to take your time driving Route 62, a 530-mile (850-kilometer) stretch of road that links Cape Town on South Africa's west coast to Port Elizabeth on the east. Not only is the journey scenic with mountain passes and canyons, but it's also dotted with incredible wineries. Spend a day or two exploring the towns along the way and sample some of the country's finest drops.

VEGAN AND VEGGIE DELIGHTS

Eating a plant-based or vegetarian diet doesn't mean you have to sacrifice flavor or fun. With so many delicious vegan desserts, vegetarian soul food, or diner classics reimagined meat-free, the world is your (plant-based) oyster.

TRY MEXICO CITY'S BEST VEGAN TACOS

Por Siempre Vegana Taquería—Mexico City, Mexico

It's always a good sign when a taco stand has dozens of locals and visitors alike crowded around eating in silence. And at Por Siempre Vegana Taquería you can't pass by without stopping to find out what the fuss is about. If you're vegan or vegetarian and worried about missing out on good tacos, worry no more: these tacos are (in our opinion) as good as or better than any you'll find in the city. They have all the usual suspects: pastor, *suadero*, steak, all made with soy or seitan. They have mushroom tacos with onion and *epazote* (an aromatic herb) and deep-fried crispy flautas stuffed with flavorful veggies. You will probably eat here more than once. Visit the stand at the corner of Calle Manzanilla and Chiapas in Roma Norte or wait for a table at their sit-down location a few blocks away.

GORGE ON YOUR FAVORITE DINER FOODS, ALL ANIMAL-FREE

The Chicago Diner—Chicago, Illinois, United States

Diner food is the ultimate comfort food; we don't make the rules. But the Chicago Diner with locations in Logan Square and Lakeview is serving up all your favorites completely meat-free since 1983. They've got vegan jalapeño poppers, hot "wings," veggie burgers, cashew cheese nachos, and their famous Reuben sandwich made with corned beef seitan and vegan Thousand Island and cheese. Are you full yet? Wash it all down with their award-winning chocolate peanut butter vegan milkshake and order a dairy-free s'mores brownie sundae for the table. You'll never want to visit a regular diner again.

ORDER (AT LEAST) ONE OF EVERYTHING AT VEGETARIAN DIM SUM

Pure Veggie House—Hong Kong, China

From the fluffy steamed barbecue "pork" buns to the *mapo* tofu to the *shao mai* stuffed with greens, Buddhist restaurant Pure Veggie House has taken everything there is to love about dim sum and turned it deliciously vegan. There's a focus on fresh, high-quality ingredients; much of the produce is sourced from local organic farms. They even have a vegan version of shark-fin soup, which is made up of glass noodles, mushrooms, and strips of bean-curd skin, and for a couple of months during winter they offer vegan hot pots. It's so hearty even carnivores won't notice there's no meat—seriously.

DEMOLISH A VEGAN DONER KEBAB

Vöner—Berlin, Germany

If you couldn't tell from the name Vöner (a combination of the words "vegan" and "döner"), this laid-back snack shop specializes in plant-based versions of the classic Turkish street food that's oh-so-popular in Berlin. The "meat" is made with a recipe of wheat protein, legumes, veggies, soy meal, and loads of herbs and spices—and grilled on a spit. It's then shaved off and stuffed into a flatbread or wrap or dished up on a plate loaded with accompaniments like fries, salads, and homemade sauces. When you're ready to come back for round two, try the vegan currywurst or seitan kebab.

SNACK YOUR HEART OUT AT THIS DAIRY-FREE CHEESE SHOP

Blue Heron—Vancouver, Canada

Vegan cheese has come a loooong way in the past few years, and the proof is in specialty plant-based creameries like Blue Heron. The vegan cheese shop produces, ages, and smokes many of its cheeses in-house—and they look and taste just like actual cheese! Sample the cashew smoke 'n' spice gouda rind washed in fermented habanero paste and smoked paprika, the beechwood-aged almond cheese, or the coconut-based bocconcini cured in herbed olive oil brine, and be prepared to mutter, "I can't believe there's no dairy in this!" over and over.

TASTE THE RAINBOW OF PLANT-BASED ETHIOPIAN FOOD

Bunna Cafe—New York, United States

This effortlessly trendy restaurant in Brooklyn's Bushwick neighborhood serves up some of the best vegan Ethiopian food going around. The game plan? Order the feast: a variety of classic Ethiopian stews and side dishes such as *gomen* (steamed collard greens with ginger and coriander), *misir wot* (red lentils cooked in a spicy berbere sauce), and *yatakilt alicha* (cabbage, carrots, potatoes, coriander, turmeric), dolloped on top of a big, round *injera*—a traditional sourdough flatbread that's tangy, spongy, and perfect for scooping up all the good stuff. Also, Bunna gets its name from the traditional Ethiopian coffee ceremony, so you'd best have a cup before you go.

VISIT THE WORLD'S OLDEST VEGETARIAN RESTAURANT

Hiltl—Zürich, Switzerland

Some cultures have been eating plant-based diets for centuries, but this restaurant in Zürich is recognized by *Guinness World Records* as the oldest vegetarian restaurant in the world. Founded in 1898, Hiltl is a fourth-generation family-run business serving up delicious homemade plant-based dishes and vegan wines. There's a by-weight or all-you-can-eat buffet section and an à la carte floor that serves plant-based goodies like truffle mushroom risotto, gnocchi with saffron sauce, or a meatless bacon cheeseburger with melty Swiss or vegan cheese. Next door, you can stock up on organic vegan meat alternatives at Hiltl Vegimetzg, Switzerland's first vegetarian butchery.

TUCK INTO DROOL-WORTHY VEGAN DONUTS

Donut Corp—Santiago, Chile

This small artisanal donut shop in Santiago whips up vegan donuts in creative, punchy flavors. Try the raspberry green tea (vanilla donut filled with raspberry mousse, covered with a matcha green tea glaze), the crème brûlée (vanilla donut filled with a soy-based pastry cream and a caramelized sugar top), or the decadent choc-nut situation (chocolate donut with a chocolate, hazelnut, and peanut butter filling and glaze). Or, hell, order a box of a dozen and try them all. You're on vacation!

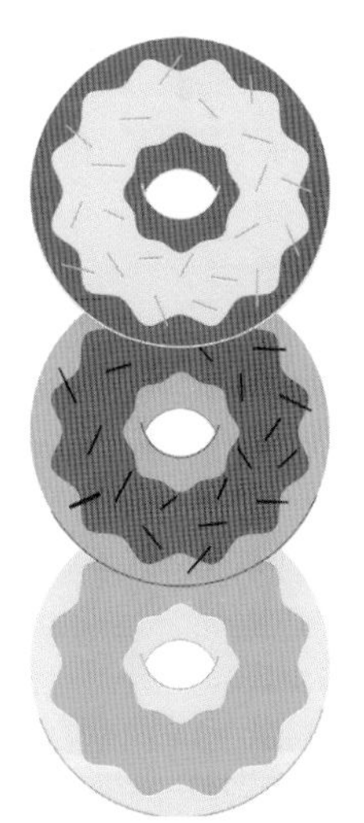

MEET YOU AT THE WORLD'S FIRST VEGAN MINI STRIP MALL

This vegan mini mall is the first of its kind in the world, and it started when four vegan businesses joined forces at the corner of Southeast Stark Street and Southeast Twelfth Avenue in Portland, Oregon, one of the world's most vegan-friendly cities. Here's who's there.

Sweet Pea Baking Co.: Come to Sweet Pea bakery for delicious vegan treats, like their heavenly chocolate peanut butter cronut.

Ice Queen: Super-cute frozen vegan treats. Ice creams, popsicles, and paletas with flavors like Watermelon Sour Patch Kids, Blueberry Pancake, and Thai Tea with Sugar Boba.

The Herbivore Clothing Co.: More than just a clothing store, Herbivore also sells accessories, gifts, and local artwork—you guessed it, all vegan.

Scapegoat Tattoo: That's right, there's a tattoo store that specializes in animal-free ink. This means that the inks and other items used don't contain any animal by-products, which they will in most other tattoo parlors.

Psst! If you're looking for Food Fight!—the all-vegan grocery store that was originally in the mini mall—check out their new location on Halsey in the Gateway District.

EAT TRADITIONAL BUDDHIST TEMPLE FOOD IN SEOUL

Sanchon—Seoul, South Korea

Step inside this traditional wooden hanok house for Seoul's best plant-based eatery, set in tranquil, plant-filled surroundings. Sanchon serves a set menu of unique appetizers and side dishes (*banchan*), all prepared with seasonal vegetables and mountain herbs. You won't leave here hungry; expect a table covered in a colorful array of simple, fresh food. The restaurant was founded by a former monk, so the menu reflects Buddhist temple cooking methods and recipes. Eating here is indeed a meditative experience and a welcome pause from all the delicious desserts and street treats you'll no doubt also be eating in Seoul.

BITE INTO VEGAN FISH AND CHIPS

Sutton & Sons Vegan—London, England

Fish and chips is a quintessential British dish—and thanks to Sutton & Sons, even those who don't eat meat can enjoy it. This family-run business has three restaurants in London, and each location serves a full vegan "fish" menu as well as traditional fish and chips (so you know the plant-based version is going to be good). To get that flaky fish texture and flavor, banana blossom is marinated in seaweed and samphire, then battered and deep fried to crispy perfection. Other meat-free items include plant-based versions of scampi, chicken, and a prawn cocktail, along with tasty sides like hand-cut chips, mushy peas, and coleslaw.

Food Tip: Remember, not every country might be familiar with the term "vegan"—but that doesn't necessarily mean there are no vegan options. A lot of countries' cuisines are traditionally heavily plant based, just without the veganism branding. Instead of asking for "vegan food," try explaining what ingredients you're trying to avoid, for example: "no meat, no eggs, no dairy."

YES, MEAT-FREE BUTCHER SHOPS ARE A THING

It's no secret that meat production has a huge impact on the environment. As more and more people move to completely plant-based diets or simply try to eat less meat, creative vegan and vegetarian butcher shops have been popping up across the globe. Here are a few of our faves.

1. **THE HERBIVOROUS BUTCHER—MINNEAPOLIS, MINNESOTA, UNITED STATES**

   This Minnesota shop specializes in small-batch, plant-based meats (think smoky ribs, chicken cutlets, and capocollo ham) and ready-to-eat sandwiches.

2. **YAM CHOPS—TORONTO, CANADA**

   At this butcher and market you'll find a well-curated selection of vegan proteins, sampler boxes, and pre-prepared meals. They also have their own range of house-made "fysh" products, which includes tunaless tuna, fysh sticks, and kalamari.

3. **SUZY SPOONS—SYDNEY, AUSTRALIA**

   Browse the selection of fourteen handmade products, including sausages, schnitzel, and meatballs, all completely vegan and preservative-free.

4. **NO BONES—SÃO PAULO AND RIO DE JANEIRO, BRAZIL**

   Developed by Chef Marcella Izzo, this animal-free butcher stocks a range of vegan meats and ready-made dishes that you can simply heat and eat.

5. **RUDY'S VEGAN BUTCHER—LONDON, ENGLAND**

   Shop the range of products at the United Kingdom's first all-plant-based butcher, then pop next door to Rudy's Diner, which serves up veganized versions of classic American diner fare, such as burgers, dairy-free milkshakes, and Reuben sandwiches.

THEMED CAFÉS AND RESTAURANTS

Dining and drinking is more fun when you're doing it somewhere with a theme. And while we toast the many memorable themed cafés, restaurants, and bars that have sadly closed their doors, here are some we're still visiting this year.

OVERLOAD ON CUTENESS AT THIS CORGI-THEMED CAFÉ

Café D.er—Alhambra, California, United States

What do corgis and soufflé pancakes both have in common? They're plump, jiggly, and both featured at Café D.er, a corgi-themed café. Come for the cute corgi butts, but stay for the soufflés—fluffy pancakes in delicious flavors like crème brûlée, tiramisu, or strawberry cheese. They also serve a huge variety of corgi-themed drinks, flavored yogurts, lunch items, and other delicious desserts. Well-behaved four-legged friends are welcomed, and when you're done taking in all this cuteness, there's a shop full of corgi items from Nayothecorgi to take home with you, too.

TAKE A SEAT AT A TOILET-THEMED DESSERT BAR

Poop Cafe—Toronto, Canada

This restaurant in Toronto's Koreatown is pure crap—in a good way. If you like potty humor and desserts, then Poop Cafe is the place for you. They serve up toilet-themed treats, including their famous freakshakes—overflowing with candy, crushed cookies, ice cream, and other deliciousness. Try the Poonicorn Freakshake topped with candy floss and a cupcake or a green tea Bing Poo served in a tiny ceramic toilet bowl. During the pandemic the café transformed into a sewing factory and started selling masks for the community and frontline workers. These days, you can sit on the throne and gorge on "toilet paper rolls"—mini crepe rolls with your filling of choice. It's all *poo*-retty great if you ask us.

SAY HELLO TO ALL THINGS MARSHMALLOW

XO Marshmallow—Chicago, Illinois, United States

This café is the childhood birthday party of dreams, a place where everything on the menu involves marshmallows. Stuff your face with gluten-free s'mores, order several desserts made with their signature gooey marshmallow fluff, or treat yourself to a Birthday Cake Hot Cocoa with cupcake syrup, sprinkles, and funfetti marshmallows floating on top. They also have a shop chock-full of dreamy treats to take home, and you can sign up for their Marshmallow Decorating Class if you want to decorate a six-by-six-inch slab of vanilla marshmallow yourself. Do or don't eat it in one sitting afterward—we would never judge!

MIX COCKTAILS IN A CAULDRON AT THIS MAGIC-THEMED BAR

The Cauldron—London, England

Calling all wizards in training: this pub in London lets you advance your magic skills by mixing your own magical potions in fantasy-themed surrounds. Potion kits are delivered to the table, and you'll be guided in making your own magic happen using molecular mixology and actual magic wands. The flagship location is in London, but there are pop-ups in other cities, including New York City. They also have a family-friendly Wands and Wizards Exploratorium in London that specializes in magical desserts and a wand-making experience. These bars and experiences are a fantasy lover's dream, but don't worry if you can't travel—they sell magic kits in their online store, too.

GET MAGICAL AT THIS UNICORN CAFÉ IN BANGKOK

Unicorn Café—Bangkok, Thailand

In case you haven't seen it on the 'gram, Bangkok's Unicorn Café is a magical pastel dreamland where you can dress up as your favorite mythical pony and order unicorn-themed desserts like rainbow crepe cake, cotton candy, or unicorn cupcakes. Even the rice and noodles are rainbow colored, and the french fries come drizzled in rainbow cheese. But the café is probably known more for its decor than its menu; hundreds of tiny unicorns hang from the ceiling, and every other surface is covered in cute unicorn and rainbow furnishings. If anyone tells you real life isn't all rainbows and unicorns, send them to this café.

IT'S ALWAYS HALLOWEEN AT THIS TIM BURTON–THEMED BAR

Beetle House—New York, New York, United States

Turn on the juice and see what shakes loose at this bar and restaurant inspired by Tim Burton, *Beetlejuice*, and all things magic and macabre, strange and unusual. Menu items range from an Edward Burger Hands half-pound burger and Sweeney Beef filet mignon to Giant Peach salad and Frankenfries. The libations are plenty and good (they have a huge range of cocktails), and the spooky atmosphere is even better. Beetle House started as a pop-up in New York City but was so popular it became permanent; they now also have a location in Los Angeles. Call his name three times, and you might even catch the ghost with the most cruising around the establishment.

HAVE A MEDIEVAL FEAST AT A VIKING RESTAURANT

Mjølner—Sydney and Melbourne, Australia

Named after Thor's hammer, Mjølner is a unique restaurant with locations in Sydney and Melbourne, Australia. The restaurant serves their modern interpretation of a Viking feast, with menu items like roasted bone marrow, venison tartare, or whiskey-glazed short rib (but there are vegetarian options, too). Start your epic feast with a shot of mead and drink beer from a horn like a true Norse person. The cocktails are inspired by Viking folklore—try the Gates of Valhalla with tequila, fenugreek syrup, and quince liqueur—and the desserts are fit for a king. You'll be dining under the protection of the god of thunder, so you're allowed to indulge.

FOLLOW YOUR NOSE TO THIS GARLIC RESTAURANT

The Stinking Rose—San Francisco and Beverly Hills, California, United States

You're not a true garlic fan until you've dined at The Stinking Rose. The garlic-themed restaurant with locations in San Francisco and Beverly Hills serves California–Italian cuisine and around three thousand *pounds* of garlic each month. And the pungent seasoning isn't limited to the lunch and dinner menus—true garlic lovers can finish their meal with Gilroy's Famous Garlic Ice Cream with chocolate sauce. Some 2,635 bulbs of garlic feature on the walls of the dining rooms, along with other quirky memorabilia and artworks. Look forward to the aroma of garlic emanating from your pores for days after a well-worth-it eating experience here.

Psst! Need more garlic? Flip back to page 143 if you missed reading about the Isle of Wight Garlic Festival.

TOKYO HAS *SO MANY* THEMED RESTAURANTS

From the Robot Restaurant, the maid cafés, and Harajuku's famous Kawaii Monster Café to all the (sometimes controversial) cat, owl, hedgehog, and bunny cafés, Tokyo loves a themed eatery. Here are three unique places to eat that you might not know about.

EAT WHAT YOU SNAG AT THIS FISHING RESTAURANT

ZAUO

They say food tastes better if you catch it yourself, and this restaurant isn't just fishing themed—it actually lets you fish for your food. You can rent a rod, snag a fish (or crustacean) from the water tanks at your feet, and then hand your catch to the staff to have it prepared however you like: as sashimi, sushi, grilled, or fried tempura style. Then, it's time to enjoy the fruits of your fishing labor in the dining area, which is aboard a big wooden boat in the center of the restaurant. There are multiple locations in Tokyo and around Japan, and some also offer sushi-making workshops for kids.

GO BACK IN TIME TO DINE IN THIS NINJA VILLAGE

NINJA AKASAKA

If you're fascinated by the Edo period in Japan—or just by ninjas—don't miss this unique ninja-themed dining and entertainment experience in Tokyo's Akasaka District (not to be confused with the better-known Asakusa). You'll wind through a secret underground path to get to your private or communal dining room, which is set in an Edo period–inspired ninja village. Ninja waiters will serve you a preset menu from acclaimed chefs, while advanced ninjas come to perform tricks for your table. You'd think with an ambience this cool the quality of food might be overlooked, but no—the food here is exceptional. This is basically fine dining but with ninjas.

GO BACK TO SCHOOL AT THIS *IZAKAYA* IN TOKYO

6年4組 (ROKUNEN YONKUMI)

With a name that translates to "sixth grade, fourth group," this *izakaya* (an informal eatery/drinking spot) is entirely school themed. With an interior made to feel like a Japanese elementary school, when you enter you'll take off your shoes and store them in a locker. Inside the classroom, you'll eat school lunch–inspired foods and drink beverages from mini "science experiments." After eating, guests are invited to do a school quiz—but good luck if you don't speak Japanese! It's a cute and fun place to stop for lunch if you want to know what it's like to go to school in Japan or if you want the weird thrill of drinking booze in a classroom.

Good to Know: Many *izakaya* and other restaurants in Japan will ask you to remove your shoes before entering. When you walk into a place, it's a good idea to scan the floors and walls for signs of other people's shoes. There are usually special slippers you can slip on when visiting the restroom, too.

ACKNOWLEDGMENTS

ILLUSTRATIONS
Jay Fleckenstein
Michael Kilian
Ivy Tai

THE TEAM AT BUZZFEED
Ines Pacheco
Emily DePaula
Eric Karp
Parker Ortolani
Richard Alan Reid
Kenneth Blom
Puja Dave
Andrea Mazey
Shira Mahler
Emma Taffet

THE TEAM AT RUNNING PRESS
Jordana Hawkins
Jess Riordan
Fred Francis
Seta Zink
Amy Cianfrone
Betsy Hulsebosch
Kara Thornton
Kristin Kiser

THE BRING ME CREW, PAST & PRESENT
Karla Agis
Daniel Arashiro
Ben Armson
Luke Bailey
Jordan Ballantine
Bibi Barud
Cheska Bacaltos
Masongsong
Maria Batalha
Kirby Beaton
Brent Bennett
Evelyn Bevans
Sumedha Bharpiliana
Sofia Munoz Boullosa
Aniket Chitnawis
Nyall Cook
Dakota Deady
Ada Enechi
Jake Frankenfield
John Gannon
Alejandra García
Andrew Gauthier
Mike Girsang
Mafer Gonzalez
Jada Harris
Kat Hartman
Liza Hicks
Nadia Honary
Auri Jackson
Vishal Jain
Phil Jahner
Kwesi James
Ashley Jones
Aishwarya Katkade
Dan Katt
Abinaya Kolanjinathan
Natalia Krslovic
Sam Linder
Justin Lee
Evelyn Liu
Tiffany Lo
Chandler Lynn
Diana Marín
Jessica Maroney
Sukanya Mathur
Skye Mayring
Omar Mendez
Ayesha Mittal
Zeta Morgan
Nancy Nguyen
Ellyse O'Halloran
Jasmine Pak
Ryan Paturzo
Kyle Richmond
Davi Rocha
Giovanni Rogel
Tania Safi
Swasti Shukla
Tom Scott
Gustavo Serrano
Jemima Skelley
Arthur Smith
Carol Tan
Jeff Thurm
Maycie Timpone
Jun Tsuboike
Emi Tulett
Lizz Warner
Quinton Washington
Julia Willing
Alessandro Zotti

. . . and everyone else who has written or produced travel content for Bring Me! and who first covered the experiences featured in this book.

SPECIAL THANKS TO:
Benjamin Juzwin
Danny Agai
Steve & Thomas Norris-Smith

Finally, thank you to all the people around the world working in the travel and hospitality industries — the small business owners, servers and bussers, gardeners and park rangers, cleaners and cashiers, airline staff and healthcare workers — and everyone else who makes the experiences in this book possible.

INDEX

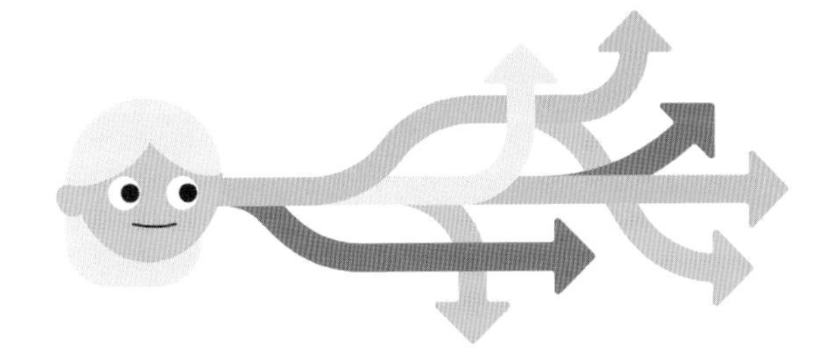

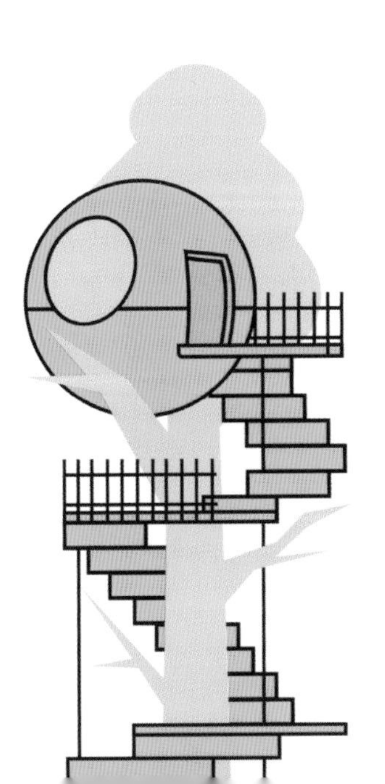

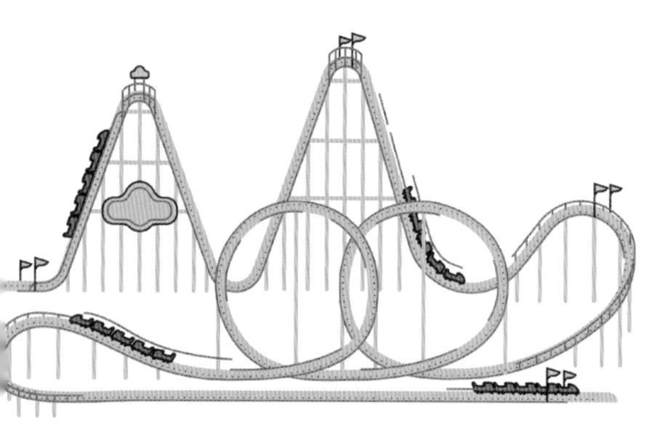

D

E

F

G

H

I

J

K

L

M

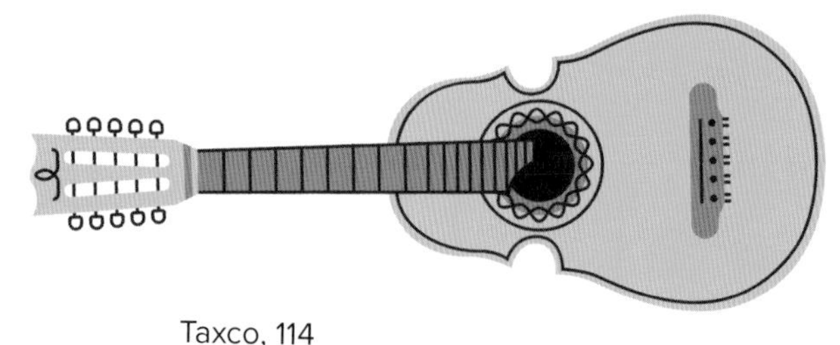

N

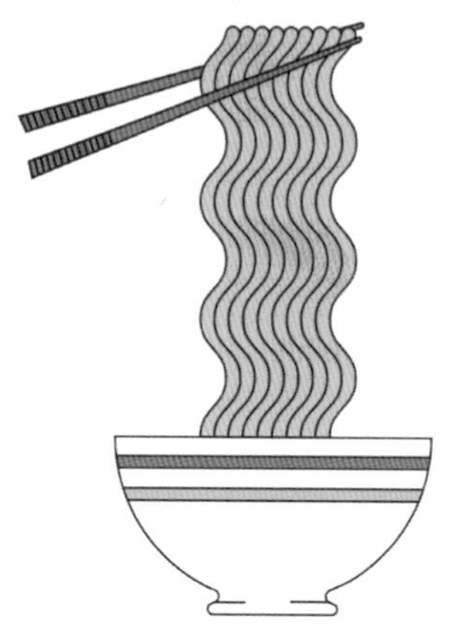

T

U

W

ABOUT THE AUTHORS

Louise Khong is an Australian writer, editor, and creative who's called Melbourne, Sydney, and New York home. She's the former director of content strategy for BuzzFeed's travel and experience brand, Bring Me. An avid traveler and advocate for paid time off, Louise has visited more than thirty countries and has touched every continent bar one (she's looking at you, Antarctica). Her favorite things to do while traveling include overpacking, learning to cook local cuisine, taking average photos of amazing places, and writing postcards but forgetting to send them.

Ayla Smith is a writer, creative, and full-time plant enthusiast living in Los Angeles with her cat, Carol. Originally from Sydney, Australia, she has been at BuzzFeed since 2014, most recently on the international team as a senior director of content. She has lived and worked in Tokyo and Mexico City and has also worked with (and out of) other BuzzFeed offices around the world. When traveling, she loves finding good nature, wandering the streets with no map, and organizing an entire trip around where and what to eat.

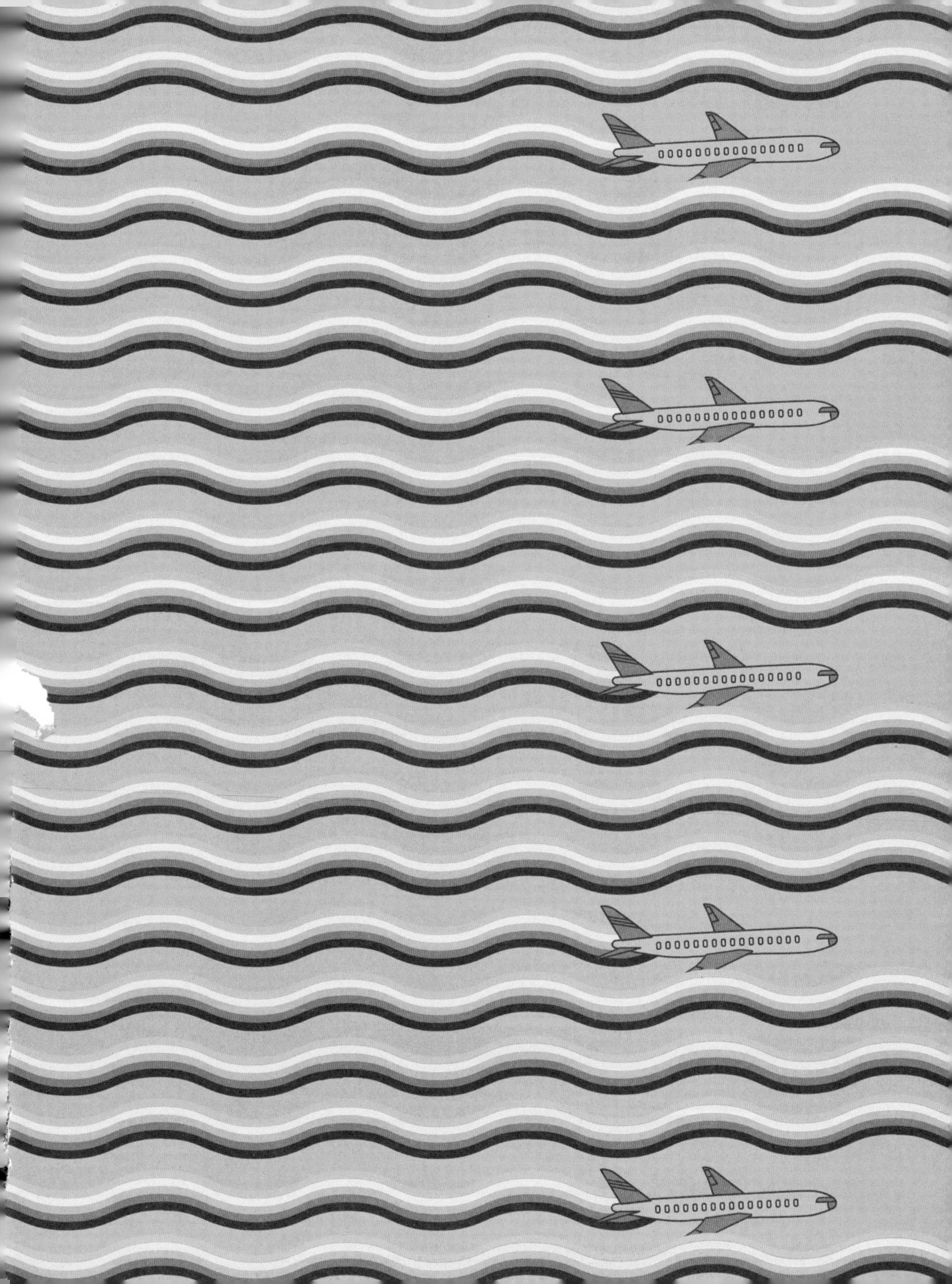